T0408047

Parasitic Plants in African Agriculture

We dedicate this work to Chris Parker, a pioneer in promoting parasitic plant research, especially in Africa, working assiduously to encourage researchers and the dissemination of knowledge on all parasitic plants to the widest possible audience. His efforts over the past six decades have helped foster a global coterie of workers in both the laboratory and the field. It is fitting that we recognize him in a book on African parasitic plants.

We also commemorate our colleague Abdul Gabbar Tayeb Babiker (1949–2022) who encouraged the production of this book, another legacy of his enthusiasm for parasitic plant research. His work in Africa inspired many younger scientists, and his collaboration with researchers across the world remains an exemplar of transnational and cross-cultural science endeavours.

Parasitic Plants in African Agriculture

Lytton John Musselman

Jonne Rodenburg

CABI is a trading name of CAB International

CABI
Nosworthy Way
Wallingford
Oxfordshire OX10 8DE
UK

Tel: +44 (0)1491 832111
E-mail: info@cabi.org
Website: www.cabi.org

CABI
200 Portland Street
Boston
MA 02114
USA

Tel: +1 (617)682-9015
E-mail: cabi-nao@cabi.org

A catalogue record for this book is available from the British Library, London, UK.

Library of Congress Cataloging-in-Publication Data

Names: Musselman, Lytton John, 1943- author. | Rodenburg, Jonne, author.
Title: Parasitic plants in African agriculture / Lytton John Musselman, Jonne Rodenburg.
Description: Wallingford, Oxfordshire; Boston, MA: CAB International, [2023] | Includes bibliographical references and index. | Summary: "Parasitic Plants in African Agriculture is the first book to describe the ecology, biology, damage caused, and control of all groups of African parasitic plants including introduced invasive species. The book is for researchers, extension workers, development officers, plant pathogists, food security specialists, weed scientists, botanists and plant ecologists"-- Provided by publisher.
Identifiers: LCCN 2023009333 (print) | LCCN 2023009334 (ebook) | ISBN 9781789247633 (hardback) | ISBN 9781789247640 (ebk) | ISBN 9781789247657 (epub)
Subjects: LCSH: Parasitic plants--Africa. | Invasive plants--Africa.
Classification: LCC SB613.A4 M87 2023 (print) | LCC SB613.A4 (ebook) | DDC 632/.52096--dc23/eng/20230602
LC record available at https://lccn.loc.gov/2023009333
LC ebook record available at https://lccn.loc.gov/2023009334

ISBN-13: 9781789247633 (hardback)
9781789247640 (ePDF)
9781789247657 (ePub)

DOI: 10.1079/9781789247657.0000

Commissioning Editor: Ward Cooper
Editorial Assistant: Emma McCann
Production Editor: Shankari Wilford

Typeset by SPi, Pondicherry, India
Printed and bound in the UK by Severn, Gloucester

Contents

About the Authors

Lytton John Musselman is Mary Payne Hogan Professor of Botany and manager of the Blackwater Ecologic Preserve in the Department of Biological Sciences at Old Dominion University, where he also served as Department Chair. His research centres on the biology of parasitic angiosperms, especially those in the Middle East and Africa. Recipient of four Fulbright awards (Sudan, West Bank, Jordan, Brunei Darussalam), he has also been a visiting professor at the American University of Beirut and the American University of Iraq, Sulaimani. He is co-founder and co-editor of *Haustorium*, the newsletter of the International Parasitic Plants Society, and served as a consultant to the International Institute of Tropical Agriculture and the International Center for Agricultural Research in the Dry Areas. His most recent books are *Edible Wild Plants of the Carolinas: A Forager's Companion* (2021) with Peter W. Schafran and *Solomon Described Plants: A Botanical Guide to Plant Life in the Bible* (2022).

Jonne Rodenburg is Professor of Agronomy at the Natural Resources Institute of the University of Greenwich, UK, where he teaches postgraduate students on weed biology and management, agronomy and crop physiology. He researches sustainable agricultural intensification and the biology, ecology and management of parasitic weeds in annual field-crop production systems in Africa. He has worked and published extensively on root-parasitic weeds, both obligate and facultative, as recipient of grants including from the Dutch Research Council NWO, the UKRI Biotechnology and Biological Sciences Research Council and The Royal Society. From 2004 to 2018, he worked at the Africa Rice Center as agronomist focusing on weed management in rice, with subsequent home bases in Mali, Benin, Senegal, Tanzania and Côte d'Ivoire. He is the current vice-president and president-elect of the International Parasitic Plants Society, subject editor at the European Weed Research Society journal *Weed Research*, published by Wiley, and associate editor at the Elsevier journal *Field Crops Research*.

Preface

More than 50 years of basic and applied research on parasitic plants has resulted in a wealth of information on their taxonomy, biology and ecology. These studies have shown the vast diversity of this fascinating group of plants in terms of species, distribution, life forms, ethnobotany and impact they have on natural and agricultural systems. Because of the latter, research endeavours have focused primarily on those species that are known as weeds. Weedy parasitic plants impact agricultural production around the world, but the African continent is probably the hardest hit. Africa harbours a wide range of different parasitic plant species, many of them endemic, and the agricultural sector in Africa is dominated by smallholder farms that are often particularly restricted in access to the knowledge and means to manage these plants when they have become weed problems. The advances in knowledge of parasitic weed biology and host–parasite interactions have resulted in valuable insights on how they could be managed to avoid crop losses. This is particularly true for some of the most widespread and damaging groups, such as the root-parasitic witchweeds and broomrapes, whereas other groups, such as stem parasites or more localized or newly emerging species, have been less studied or are overlooked. Importantly, for even the most widely studied species such as the witchweeds, African smallholder farmers have benefited little from scientific advances and insights and the control technologies resulting from these. This book aims to compile what is known so far about the parasitic weeds on this continent, and what is not. It further tries to answer questions such as 'which species are important and where?', 'which species are expected to increase?', 'how can they be managed?' and 'why is it so difficult to solve parasitic weed problems?'.

Acknowledgements

We thank Chris Randle, Daniel Nickrent, Juma Kayeke, Kamal I. Mohamed, Kushan Tennakoon, Oumar Ouédraogo and Yaacov (Coby) Goldwasser for their critical review of sections of this book.

Several of our colleagues gave approval for use of photographs; these are Chris Thorogood, Emmanuel I. Aigbokhan, Harro J. Bouwmeester, Mamadou Cissoko and Steven M. Runo. The remaining images are by the authors. Some images were in ECHO Technical Note 94 and are used here by permission.

Justin F. Djagba of AfricaRice is kindly acknowledged for his help in generating species distribution maps provided in this volume.

1 Introduction to Parasitic Plants

Abstract

Parasitic plants are generally little known by agriculturalists even though under some conditions they may be the most important factor in crop losses. Globally, they have their greatest impact on food crops of smallholder farming systems in Africa. Despite this, they are often treated simply as weeds, overlooking the fact that they do not just harm the crop indirectly through competition, but also directly through parasitism. Parasitic weeds connect to crop plants through a specialized structure, the haustorium. In fact, the presence of a haustorium is what defines a parasitic plant. Parasitism has arisen in 12 clades of angiosperms, yielding plants with a diversity of habits including herbs, vines, shrubs and even trees. Likewise, there is a range of parasite–host interactions. Some parasites will only germinate with a stimulant produced by the host. Some are specific in host selection, some are promiscuous with many different hosts, and some are not quite generalists but are not host specific. We include all known African parasites that attack crops, with emphasis on mistletoes, witchweeds, dodders and broomrapes, including their taxonomy, hosts, distribution and control measures.

1.1 Parasitic Plants as Weed Problems

Parasitism has been reported in 28 plant families, comprising nearly 4500 species, all exclusively dicotyledons (Heide-Jørgensen, 2013; Nickrent, 2020). When these plants parasitize other plants, either out of necessity or to increase their reproductive output (Shen *et al.*, 2006), the host plants can be severely damaged. When hosts are agricultural crops, parasitic plants can become important weed problems. A broad range of African crops suffer from parasitic weeds. Affected crops include staple food grains (e.g. maize, rice, sorghum, millet) and legumes (e.g. cowpea, faba beans, lentils), a diversity of vegetables (e.g. carrots, tomatoes, leek), oil crops (e.g. sunflower, linseed), fibre crops (e.g. flax, hemp), forage crops (e.g. lucerne, clover), many fruit-tree species (e.g. mango, guava, citrus) and plantation cash crops (e.g. cacao, coffee, tea, rubber).

(L.J. Musselman and J. Rodenburg)

DOI: 10.1079/9781789247657.0001

Parasitic plants can lead to severe yield losses, making them an important constraint to food security in many areas (Fig. 1.1). While quantitative information on yield losses from parasitic weeds is lacking for many parasite–host species combinations, available data emphasize just how serious these pathogens are. An assessment by Rodenburg *et al.* (2016a) showed that when *Striga asiatica* is not controlled, mean yield losses of upland rice are around 73%. For maize, the same parasite causes yield losses of 80% or higher when uncontrolled (Ransom *et al.*, 1990; Rusinamhodzi *et al.*, 2012). *Striga hermonthica* can cause yield losses of up to 84% (mean: 37%) in sorghum (Rodenburg *et al.*, 2005) and up to 81% in maize (mean: 68%; Kim *et al.*, 2002), depending on variety, infestation level and environmental conditions. *Rhamphicarpa fistulosa* causes yield losses of rice ranging from 24% to 73% (mean: 50%), again depending on the variety and infestation level (Rodenburg *et al.*, 2016b). Field dodder, *Cuscuta campestris*, reduces yields of sesame by 67%, soybean by 48%, pigeon pea by 25% and groundnut by 18% (Mishra *et al.*, 2007). *Alectra vogelii* inflicted yield losses in susceptible cowpea varieties that were reported to range from 30% to 66% (mean: 51%; Alonge *et al.*, 2001) whereas *Striga gesnerioides* inflicted yield losses that ranged from 79% to 86% (mean: 81%; Alonge *et al.*, 2005), but for both parasite species, these losses were reduced in some of the resistant and tolerant cowpea genotypes. Yield losses caused by broomrapes (*Orobanche* spp. and *Phelipanche* spp.) in faba bean, chickpea, tomato, potato and sunflower range from 5% to 100% (Abang *et al.*, 2007).

No quantitative data exist on damage inflicted by mistletoe but assessment is based on field observations and farmer perceptions. Loranthaceae parasitism causes important shea tree yield reductions in Burkina Faso (Boussim *et al.*, 2004). Damage to these and other economically important crops is generally increased by low soil fertility and drought stress, conditions facing many African smallholders.

The above yield-loss estimates are field- or crop-scale measurements. The extent of the parasitic weed problem in Africa cannot be truly assessed without quantitative information on the spread of the different parasite species across croplands and their economic impact at a national and regional scale. The data on parasitic weed distribution and economic impact in Africa are scarce, however, and are mainly associated with those parasites that impact the region's cereal production. Maize cropland infested by *Striga* spp. (chiefly *Striga hermonthica* and *S. asiatica*) in sub-Saharan Africa is estimated at 2.3 million ha and the concomitant annual economic losses are estimated at US$383 million (Woomer *et al.*, 2008). The area of rainfed rice infested by parasitic weeds (*Striga hermonthica, S. asiatica, S. aspera* or *Rhamphicarpa fistulosa*) is estimated at 1.34 million ha (about 19% of the total area under rainfed rice) resulting in a total estimated annual economic impact of at least US$111 million (Rodenburg *et al.*, 2016a). The total annual loss caused by *S. hermonthica*, one of the main parasitic weeds in cereals in Africa, is roughly estimated to be more than US$1 billion (Parker, 2009). For Africa, no quantitative economic impact data are available on any of the stem parasites described in this book.

Fig. 1.1. Farmers in parasitic-weed-infested field crops in Africa. (A) Rice field infested by *Striga hermonthica* (purple-flowered plants) in Côte d'Ivoire. (B) Rice field infested by *Rhamphicarpa fistulosa* (reddish plants among the rice) in Uganda.

1.2 What is a Parasitic Plant?

Although parasitic plants are often thought of as weeds, they are part of a guild of highly unique plants, the parasitic angiosperms. An understanding of their biology is essential for effective control and management. Parasitic plants are amazingly specialized, with remarkable adaptations for their heterotrophic existence. Their habits are diverse, including herbaceous plants, vines, shrubs and trees. Some appear innocuous, with no external evidence of their parasitic nature. Others lack chlorophyll or even leaves and stems, existing only within the bodies of other plants until they flower. Parasitic plants' reproductive strategies also vary widely, from the tiny (1 mm) flowers of some mistletoes to the metre-wide flowers of *Rafflesia* species – the largest flower in the world. Unique among African parasitic plants is the rainforest tree *Okoubaka aubrevillei* (Santalaceae), a rare but widely distributed tree in Western and Central Africa, much sought after for its purported medicinal value. It is the largest parasitic plant in the world and little studied. Veenendaal *et al.* (1996) present the only data from experimental work on host selection and host damage. In their study, they found that *O. aubrevillei* caused morbidity and death in seedlings of *Pericopsis elata*, a leguminous rainforest tree. The authors suggest that *O. aubrevillei* favours such nitrogen-fixing trees and that the role of parasitism is to reduce competition at the seedling stage.

What this diverse coterie of plants share is a haustorium. Simply put, if a haustorium is present, the plant is a parasite. It is the defining feature of this group of organisms. The haustorium is the morphological and physiological bridge between host and parasite. This structure is the conduit for water and dissolved materials, such as nutrients and metabolites, but also proteins and pathogens (Yoshida *et al.*, 2016) as well as genetic material transported from the host into the parasite or from the parasite into the host. Non-parasitic weeds compete with crop plants for water and nutrients in the soil, whereas parasitic weeds obtain these resources directly from host plants. Farmers are sometimes surprised to learn that some of the weeds in their crops, in particular the ones with green leaves such as witchweeds, are also parasites. Knowing the parasitic behaviour is, however, essential to understanding control measures.

1.3 Categories of Parasitic Weeds

There are roughly four different categories of parasitic plants (Table 1.1). Parasitic plants can be distinguished by the presence or absence of chlorophyll. Those that produce chlorophyll (and therefore have some photosynthetic activity) are termed hemiparasites (also known as semiparasites), and this category comprises about 90% of all parasitic plant species (Heide-Jørgensen, 2013). Those that lack chlorophyll (and therefore are not green and are totally dependent upon their host for nutrition and water) are termed holoparasites. Another distinction among parasites is with germination. Obligate parasites require the presence of a host to germinate and initiate a haustorium. Facultative parasites, on the other hand, can germinate without a host (see Kabiri *et al.*, 2016).

Table 1.1. Parasitic plant species reported to be weed problems in African agriculture.

Parasitism and common name	Family	Genus	Species	Main crop hosts	Chapter[a]
Stem parasites					
Obligate hemiparasites					
Mistletoe	Loranthaceae				
		Tapinanthus	*T. bangwensis*	Guava, other tree crops	2
			T. belvisii	Tree crops	2
		Erianthemum	*E. dregei*	Tree crops	2
		Phragmanthera	*P. capitata*	Tree crops	2
			P. incana		2
	Viscaceae				
		Viscum	*V. cruciatum*	Tree crops	2
			V. anceps		2
			V. engleri		2
			V. rotundifolium		2
Love vine	Lauraceae				
		Cassytha	*C. filiformis*	Mango, cashew, other tree crops	3
Dodder	Convolvulaceae				
Field dodder		***Cuscuta***	***C. campestris***	Vegetables, tree crops, forage crops	4
			C. epilinum	Flax	4
Fringed dodder			*C. suaveolens*		4
			C. hyalina		4
Eastern dodder			*C. monogyna*	Tree crops	4
Lucerne dodder			*C. epithymum*	Forage crops, lucerne	4
			C. pedicellata	Vetch, lentil, arugula	4
			C. planiflora	Lucerne	4
Australian dodder			*C. australis*	Lucerne	4
Chinese dodder			*C. chinensis*	Various	4
			C. kilimanjari	Cassava, ornamental shrubs	4

Continued

Table 1.1. Continued.

Parasitism and common name	Family	Genus	Species	Main crop hosts	Chapter[a]
Root parasites					
Facultative hemiparasites					
Rice vampire weed	Orobanchaceae				
		Rhamphicarpa	***R. fistulosa***	Rice, maize	5
			R. brevipedicellata		5
			R. capillacea		5
			R. elongata		5
			R. veronicaefolia		5
Hairy buchnera	Orobanchaceae				
		Buchnera	*B. hispida*	Cereals	6
NA	Orobanchaceae				
		Micrargeria	*M. filiformis*	Rice	11
		Sopubia	*S. parviflora*	Rice	11
Thesium	Santalaceae/ Thesiaceae				
		Thesium	*T. humile*		11
			T. resedoides		11
Obligate hemiparasites					
Witchweed	Orobanchaceae				
Red witchweed		***Striga***	***S. asiatica***	Sorghum, millet, maize, rice, sugarcane	7
Witchweed			*S. aspera*	Cereals, fonio, sugarcane	7
Purple witchweed			***S. hermonthica***	Sorghum, millet, maize, rice, sugarcane	7
Cowpea witchweed			***S. gesnerioides***	Cowpea, sweet potato, tobacco	7
Giant maize witchweed			*S. forbesii*	Cereals, sugarcane	7
NA			*S. passargei*	Cereals	7
NA			*S. brachycalyx*	Cereals	7
Alectra	Orobanchaceae				

		Alectra	***A. vogelii***	Cowpea, groundnut	8
			A. picta	Pulses	8
			A. sessiliflora	Pulses	8
			A. orobanchoides	Sunflower, tobacco	8
Obligate holoparasites					
Broomrape	Orobanchaceae				
Bean broomrape		*Orobanche*	*O. crenata*	Faba bean, bean, chickpea	9
Sunflower broomrape			***O. cumana*** syn. ***cernua***	Sunflower, tobacco, tomato, aubergine	9
Small broomrape			*O. minor*	Tobacco, lettuce, forage legumes	9
Stinking broomrape			*O. foetida*	Pulses	9
Egyptian broomrape		*Phelipanche*	*P. aegyptiaca*	Potato, tomato, melons	9
Branched broomrape			***P. ramosa***	Potato, tomato, aubergine, tobacco, cole crops	9
Thonningia	Balanophoraceae				
		Thonningia	*T. sanguinea*	Tree crops including rubber, coffee and cacao	10

Species in bold are the most economically important.
[a]Chapter in this volume where the species is discussed.
NA = no widely accepted common name available.

Intuitively, it seems that the most serious parasitic weeds would be holoparasites. And indeed, species of *Orobanche* and *Phelipanche* are well-known pathogens of a variety of crops. But in Africa, the most serious parasitic weeds are the witchweeds, which are obligate hemiparasites in the genus *Striga*. A further broad distinction can be made between categories of parasitic weeds in terms of where they parasitize their hosts. Around 40% of parasitic plants attack stems, whereas others are restricted to roots. These are simply referred to respectively as stem parasites and root parasites.

1.4 Parasitic Plant Research

The modern science of parasitic plants was launched in 1969 by the publication of Job Kuijt's magisterial biology of parasitic plants (Kuijt, 1969). This drew attention to a group of plants known chiefly for their bizarre morphology. A decade earlier, in-depth studies on physiology, biochemistry and control were stimulated by the discovery of *Striga asiatica* (red witchweed) in North and South Carolina (USA) in the 1950s. The parasite quickly developed as a serious pathogen of maize in these states, prompting extensive work on the biology, control and containment of this species. As a result, after many years of work, the elegant, complex germination biology of witchweed and other parasites has been elucidated and parasitic plant research expanded worldwide, leading to a surge in publications on parasitic plants.

Following Kuijt's treatment, a series of books on parasitic plants has appeared, for example Parker and Riches (1993), Press and Graves (1995), Heide-Jørgensen (2008) and Joel *et al.* (2013). These volumes deal with parasitic angiosperms as a whole. Less exhaustive discussions of parasitic plants are summarized in Těšitel (2016), Nickrent and Musselman (2017), and Texeira-Costa and Davis (2021). Reviews of groups of parasites we cover in this book can be found in their respective chapters.

As the number of publications suggests, an appraisal and review of research would be a large undertaking, beyond the scope of this book. As examples, two highlights stand out: first, phylogenetic studies, well reviewed in Nickrent (2020) documenting the evolution of parasitism in 12 clades of angiosperms; and second, the germination biology of parasites, especially root parasites. This has resulted in the discovery of a new group of plant hormones, the strigolactones. These growth regulators are now known to be widespread in angiosperms. A helpful review of strigolactones is provided by Xie *et al.* (2010).

The heightened level of research in parasitic plants is now a worldwide phenomenon. In 1957, in response to the discovery of witchweed in the USA, an exhaustive review of the world literature on witchweed was published as a detailed annotated bibliography (McGrath *et al.*, 1957). It had 298 references, including non-peer-reviewed entries. A November 2022 Web of Science search (all peer reviewed) for *Striga* yielded 1801 strikes. Similarly, an extensive review of *Cuscuta* in 1994 (Dawson *et al.*, 1994) had 303 references, the Web of Science search for *Cuscuta* gave 1271, and for *Orobanche* (including *Phelpanche*) (broomrapes) about 1585. Of course, not all the references concern agriculture or even biology. Studies on these plants have expanded beyond agronomic interest to phylogenetic research, physiology, herbal medicines, ecology and more.

Despite the thousands of studies by scientists around the world, smallholder farmers in Africa have profited little by the effort and expense put into understanding parasitic weeds. Control, either by reducing infestations or by reducing the impact on the host, is seldom realized by the farmer whose management of the parasites affects daily existence. It has been previously observed by Schut *et al.* (2015a) that research on parasitic weeds in Africa has mainly focused on understanding the biology, ecology and distribution of the parasites, and on the development and testing of strategies for managing them, with some efforts on understanding the socio-cultural dimension (e.g. Vissoh *et al.*, 2008; N'cho *et al.*, 2014) and economic impact of parasitic weeds (e.g. N'cho *et al.*, 2017, 2019). The institutional and political dimensions of parasitic weeds and the innovations to address them have not received the same structural attention. While farmers frequently participate in parasitic weed research (e.g. Schulz *et al.*, 2003; Emechebe *et al.*, 2004; Abang *et al.*, 2007; Tippe *et al.*, 2017), the private sector, civil society organizations and government representatives are less often involved (Schut *et al.*, 2015b). For research on parasitic weeds to benefit smallholder farmers, involving a broader range of stakeholders and considering broader dimensions than just the crop or farm is deemed necessary (Rodenburg *et al.*, 2015).

1.5 Parasitic Weeds in African Agricultural Systems

Parasitic weed infestation, in particular by species of the Orobanchaceae, constitutes one of the most important and complex agricultural production constraints in Africa (e.g. Vurro *et al.*, 2010; Waddington *et al.*, 2010). The problem is important because staple crops such as maize, rice, sorghum and millet are important hosts of a number of the parasitic weed species (e.g. *Striga hermonthica*, *S. asiatica*, *Rhamphicarpa fistulosa*) and because these species are widely distributed (e.g. Rodenburg *et al.*, 2016b). Hence, the parasitic weed problem greatly affects food security in the region.

The problem is complex because of the ingenious biology of plant parasitism (see Shen *et al.*, 2006; Spallek *et al.*, 2013; Těšitel, 2016). Many weedy species of parasitic plants have a wide host range, and their germination and reproductive biology render them highly successful in annually cropped environments. The problem is also difficult because most of the affected crops in Africa are predominantly grown by smallholder farmers. Although smallholder farming systems in Africa are highly diverse in their resources, environments, challenges and opportunities (e.g. Tittonell *et al.*, 2010), the majority of farmers struggle with adverse environmental conditions and limited access to productive agricultural land, production resources, information and services. These conditions render the control of parasitic weeds an even more difficult task.

Parasitic weed infection and damage is often associated with and aggravated by adverse biophysical conditions such as poor soil fertility and drought. The weeds present technological challenges because the number of feasible, effective and affordable control measures is limited (e.g. Tippe *et al.*, 2017; Silberg *et al.*, 2020) or farmers are unaware of them. The affordability, accessibility and awareness of control strategies are a direct function of the socio-cultural, economic, institutional and even political dimensions shaping this problem; agricultural extension services in rural Africa are often poorly staffed, poorly equipped and ill-informed on parasitic weed problems and ways to address them, and communications between farmers and extension and crop protection services are often suboptimal (Schut *et al.*, 2015b). Therefore, addressing the problem of parasitic weeds in Africa, by technological and organizational control strategies, requires not only a thorough understanding of the biology and ecology of the important species but also a better understanding of the social, economic and institutional environments where these weeds are problems. Such research and development endeavours need to involve a range of stakeholders, including social and natural science researchers, farmers, extension services, and public and private crop health services. The control strategies arising from such a transdisciplinary research approach should match the resource availability and farming practices of the farmers who need to implement them and should be effectively communicated to them and be locally available at an affordable price or input level.

The present work deals with parasitic plants that are current or potential agricultural pests (Table 1.1). Although it is beyond the scope of this book to note them all, parasitic species that are not currently a problem in Africa possess – at least theoretically – the ability to become weedy and cause crop damage in the future. There are examples of indigenous parasitic plants becoming pathogens in agriculture and forestry (e.g. *Thonningia sanguinea* on rubber, coffee and other crops in Western Africa; Imarhiagbe and Aigbokhan, 2019). Knowledge on biology and control of those species representing current parasitic weed problems in Africa, as well as on the socio-economic and institutional environments of farming systems where these problems are embedded, could prepare us for future outbreaks. The hope of the authors is that this contribution will increase the awareness of these plants as parasitic pathogens – especially those that are currently lesser known – ultimately to aid the smallholder farmer in Africa.

References

Abang, M.M., Bayaa, B., Abu-Irmaileh, B. and Yahyaui, A. (2007) A participatory farming system approach for sustainable broomrape (*Orobanche* spp.) management in the Near East and North Africa. *Crop Protection* 26, 1723–1732.

Alonge, S.O., Lagoke, S.T.O. and Ajakaiye, C.O. (2001) Cowpea reactions to *Alectra vogelii* II: effect on yield and nutrient composition. *Crop Protection* 20, 291–296.

Alonge, S.O., Lagoke, S.T.O. and Ajakaiye, C.O. (2005) Cowpea reactions to *Striga gesnerioides* II. Effect on grain yield and nutrient composition. *Crop Protection* 24, 575–580.

Boussim, I.J., Guinko, S., Tuquet, C. and Sallé, G. (2004) Mistletoes of the agroforestry parklands of Burkina Faso. *Agroforestry Systems* 60, 39–49.

Dawson, J.H., Musselman, L.J., Wolswinkel, P. and Dörr, I. (1994) Biology and control of *Cuscuta*. *Reviews of Weed Science* 6, 265–317.

Emechebe, A.M., Ellis Jones, J., Schulz, S., Chikoye, D., Douthwaite, B. *et al.* (2004) Farmers' perception of the *Striga* problem and its control in Northern Nigeria. *Experimental Agriculture* 40, 215–232.

Heide-Jørgensen, H.S. (2008) *Parasitic Flowering Plants*. Brill, Leiden, The Netherlands.

Heide-Jørgensen, H.S. (2013) Introduction: the parasitic syndrome in higher plants. In: Joel D.M., Gressel J. and Musselman, L.J. (eds) *Parasitic Orobanchaceae*. Springer, Berlin, pp. 1–18.

Imarhiagbe, O. and Aigbokhan, E.I. (2019) Studies on *Thonningia sanguinea* Vahl (Balanophoraceae) in southern Nigeria. Range and host preference. *International Journal of Conservation Science* 10, 721–732.

Joel, D.M., Gressel, J. and Musselman, L.J. (eds) (2013) *Parasitic Orobanchaceae: Parasitic Mechanisms and Control Strategies.* Springer, Berlin.

Kabiri, S., Van Ast, A., Rodenburg, J. and Bastiaans, L. (2016) Host influence on germination and reproduction of the facultative hemi-parasitic weed *Rhamphicarpa fistulosa*. *Annals of Applied Biology* 169, 144–154.

Kim, S.K., Adetimirin, V.O., The, C. and Dossou, R. (2002) Yield losses in maize due to *Striga hermonthica* in West and Central Africa. *International Journal of Pest Management* 48, 211–217.

Kuijt, J. (1969) *The Biology of Parasitic Flowering Plants.* University of California Press, Berkeley, California.

McGrath, H., Shaw, W.C., Jansen, L.L., Lipscomb, B.R., Miller, P.R. *et al.* (1957) *Witchweed (*Striga asiatica*) – A New Parasitic Plant in the United States.* US Department of Agriculture, Special Publication 10, Washington, DC.

Mishra, J.S., Moorthy, B.T.S., Bhan, M. and Yaduraju, N.T. (2007) Relative tolerance of rainy season crops to field dodder (*Cuscuta campestris*) and its management in niger (*Guizotia abyssinica*). *Crop Protection* 26, 625–629.

N'Cho, S.A., Mourits, M., Rodenburg, J., Demont, M. and Lansink, A.O. (2014) Determinants of parasitic weed infestation in rainfed lowland rice in Benin. *Agricultural Systems* 130, 105–115.

N'Cho, S.A., Mourits, M., Demont, M., Adegbola, P.Y. and Lansink, A.O. (2017) Impact of infestation by parasitic weeds on rice farmers' productivity and technical efficiency in sub-Saharan Africa. *African Journal of Agricultural and Resource Economics* 12, 35–50.

N'Cho, S.A., Mourits, M., Rodenburg, J. and Lansink, A.O. (2019) Inefficiency of manual weeding in rainfed rice systems affected by parasitic weeds. *Agricultural Economics* 50, 151–163.

Nickrent, D.L. (2020) Parasitic angiosperms: how often and how many? *Taxon* 69, 5–27.

Nickrent, D.L. and Musselman, L.J. (2017) Parasitic plants. In: Ownley, B.H. and Trigiano, R.N. (eds) *Plant Pathology: Concepts and Laboratory Exercises, 3rd edn.* CRC Press, Boca Raton, Florida, pp. 277–288.

Parker, C. (2009) Observations on the current status of *Orobanche* and *Striga* problems worldwide. *Pest Management Science* 65, 453–459.

Parker, C. and Riches, C.R. (1993) *Parasitic Weeds of the World*. CAB International, Wallingford, UK.

Press, M.C. and Graves, J.D. (eds) (1995) *Parasitic Plants*. Chapman and Hall, London.

Ransom, J.K., Eplee, R.E. and Langston, M.A. (1990) Genetic variability for resistance to *Striga asiatica* in maize. *Cereal Research Communications* 18, 329–334.

Rodenburg, J., Bastiaans, L., Weltzien, E. and Hess, D.E. (2005) How can field selection for *Striga* resistance and tolerance in sorghum be improved? *Field Crops Research* 93, 34–50.

Rodenburg, J., Schut, M., Demont, M., Klerkx, L., Gbehounou, G. *et al.* (2015) Systems approaches to innovation in pest management: reflections and lessons learned from an integrated research program on parasitic weeds in rice. *International Journal of Pest Management* 61, 329–339.

Rodenburg, J., Cissoko, M., Dieng, I., Kayeke, J. and Bastiaans, L. (2016a) Rice yields under *Rhamphicarpa fistulosa*-infested field conditions, and variety selection criteria for resistance and tolerance. *Field Crops Research* 194, 21–30.

Rodenburg, J., Demont, M., Zwart, S.J. and Bastiaans, L. (2016b) Parasitic weed incidence and related economic losses in rice in Africa. *Agriculture, Ecosystems & Environment* 235, 306–317.

Rusinamhodzi, L., Corbeels, M., Nyamangara, J. and Giller, K.E. (2012) Maize–grain legume intercropping is an attractive option for ecological intensification that reduces climatic risk for smallholder farmers in central Mozambique. *Field Crops Research* 136, 12–22.

Schulz, S., Hussaini, M.A., Kling, J.G., Berner, D.K. and Ikie, F.O. (2003) Evaluation of integrated *Striga hermonthica* control technologies under farmer management. *Experimental Agriculture* 39, 99–108.

Schut, M., Rodenburg, J., Klerkx, L., Kayeke, J., Van Ast, A. *et al.* (2015a) RAAIS: Rapid Appraisal of Agricultural Innovation Systems (Part II). Integrated analysis of parasitic weed problems in rice in Tanzania. *Agricultural Systems* 132, 12–24.

Schut, M., Rodenburg, J., Klerkx, L., Hinnou, L.C., Kayeke, J. *et al.* (2015b) Participatory appraisal of institutional and political constraints and opportunities for innovation to address parasitic weeds in rice. *Crop Protection* 74, 158–170.

Shen, H., Ye, W., Hong, L., Huang, H., Wang, Z. *et al.* (2006) Progress in parasitic plant biology: host selection and nutrient transfer. *Plant Biology* 8, 175–185.

Silberg, T.R., Richardson, R.B. and Lopez, M.C. (2020) Maize farmer preferences for intercropping systems to reduce *Striga* in Malawi. *Food Security* 12, 269–283.

Spallek, T., Mutuku, M. and Shirasu, K. (2013) The genus *Striga*: a witch profile. *Molecular Plant Pathology* 14, 861–869.

Těšitel, J. (2016) Functional biology of parasitic plants: a review. *Plant Ecology and Evolution* 149, 5–20.

Texeira-Costa, L. and Davis, C.C. (2021) Life history, diversity, and distribution of parasitic flowering plants. *Plant Physiology* 187, 32–51.

Tippe, D.E., Rodenburg, J., Schut, M., Van Ast, A., Kayeke, J. *et al.* (2017) Farmers' knowledge, use and preferences of parasitic weed management strategies in rain-fed rice production systems. *Crop Protection* 99, 93–107.

Tittonell, P., Muriuki, A., Shepherd, K.D., Mugendi, D., Kaizzi, K.C. *et al.* (2010) The diversity of rural livelihoods and their influence on soil fertility in agricultural systems of East Africa – a typology of smallholder farms. *Agricultural Systems* 103, 83–97.

Veenendaal, E.M., Aberese, I.K., Walsh, M.F. and Swaine, M.D. (1996) Root parasitism in a West African rainforest tree *Okoubaka aubrevillei* (Santalaceae). *New Phytologist* 134, 487–493.

Vissoh, P.V., Gbèhounou, G., Ahanchede, A., Roling, N.G. and Kuyper, T.W. (2008) Evaluation of integrated crop management strategies employed to cope with *Striga* infestation in permanent land use systems in southern Benin. *International Journal of Pest Management* 54, 197–206.

Vurro, M., Bonciani, B. and Vannacci, G. (2010) Emerging infectious diseases of crop plants in developing countries: impact on agriculture and socio-economic consequences. *Food Security* 2, 113–132.

Waddington, S.R., Li, X.Y., Dixon, J., Hyman, G. and de Vicente, M.C. (2010) Getting the focus right: production constraints for six major food crops in Asian and African farming systems. *Food Security* 2, 27–48.

Woomer, P.L., Bokanga, M. and Odhiambo, G.D. (2008) *Striga* management and the African farmer. *Outlook on Agriculture* 37, 277–282.

Xie, S., Yoneyama, K. and Yoneyama, K. (2010) The strigolactone story. *Annual Review of Phytopathology* 48, 93–117.

Yoshida, S., Cui, S.K., Ichihashi, Y. and Shirasu, K. (2016) The haustorium, a specialized invasive organ in parasitic plants. *Annual Review of Plant Biology* 67, 643–667.

Part I Stem Parasites

Stem parasites attach to the stem portion of the host. They are hemiparasites, containing chlorophyll. These plants are also obligate parasites. Although they can germinate without a stimulant, they are still obligate because they cannot mature without attaching to a host.

Important weedy representatives of stem parasites include the three groups presented in the following chapters: mistletoes (Chapter 2), the largest in terms of species; love vines (Chapter 3), *Cassytha filiformis*, an under-studied parasite; and dodders (Chapter 4), species of the genus *Cuscuta*. Dodders comprise a large group of parasitic angiosperms and some of them, such as *C. campestris*, combine a very broad distribution with a wide host plant range. Dodders are therefore the most serious stem-parasitic weeds of field crops in Africa. Mistletoes, also little studied, attack only woody plants and are very difficult to control.

2 Mistletoe

Abstract
Mistletoes are often the most conspicuous parasites in the flora. They are shrubs that grow on woody hosts. Some mistletoes are obligate parasites of other mistletoes. Crops damaged by mistletoes include rubber and various fruit trees. The haustorium of the mistletoe forms an intimate graft with its hosts, making control of the parasite harmful to the host. Control is usually by excising the mistletoe from the host tree, a difficult and expensive method. Even if the parasite is removed, long strands of tissue can extend far beyond the point of attachment of the shrub. These tentacles can then produce new shrubs. African mistletoes pose a potential threat when tree crops are planted in their vicinity and the lack of knowledge about effective and acceptable control measures necessitates much additional study.

2.1 Introduction

Mistletoes are parasitic flowering plants in the sandalwood order (Santalales) that attach to the stem of other plants (definition adapted from Mathiasen *et al.*, 2008). In Africa all mistletoes are epiphytic shrubs on host trees and contain chlorophyll, although the dwarf mistletoes are dependent upon their hosts to augment their photosynthesis. Mistletoes are the largest group of parasitic angiosperms on the continent, parasitizing a greater diversity of host families than any other parasites. Despite their number of species and their range throughout most of Africa, they have received less attention than root-parasitic weeds such as species of *Striga* and *Orobanche*. For the general biology of mistletoes see Kuijt (1969), Calder and Bernhardt (1984), Parker and Riches (1993), Mathiasen *et al.* (2008) and Heide-Jørgensen (2008).

The two families of African mistletoes, Loranthaceae and Viscaceae, their structure, evolution, distribution and ecology are discussed in Polhill and Wiens (1998), who describe 47 species of Viscaceae and 237 of Loranthaceae. Visser (1981), although limited to Southern Africa, includes images of mistletoes

(L.J. Musselman and J. Rodenburg)
DOI: 10.1079/9781789247657.0002

and information on their life cycle. The phylogeny of the mistletoes has been clarified through extensive molecular studies summarized in Nickrent (2020). Taxonomic studies have resulted in numerous synonyms that are not covered here. The purpose of the treatment in this volume is to provide a synopsis of the diversity, biology, hosts and control of these fascinating plants in their relationship to African agriculture.

2.2 Biology

All mistletoes in Africa are evergreen (Glatzel *et al.*, 2017) even if the host is deciduous, an adaptation favouring retention of nutrients. Leaves, when present, are always simple and have entire margins. The leaf is the photosynthetic organ and can mimic the leaves of the host (e.g. Polhill and Wiens, 1998). Flowers of the two families are quite different; those of Viscaceae are small (Fig. 2.1). Floral biology in members of the Viscaceae has received scant attention. Flowers in Loranthaceae, on the other hand, are characteristically colourful and adapted to bird pollination (Fig. 2.2).

The parasitic organ of mistletoes, the haustorium, varies among genera (Fig. 2.3). Haustorial morphology has been a topic of interest for some time, with a review by Calvin and Wilson (1998). A more recent paper by Teixeira-Costa *et al.* (2020) reviews the development and morphology of mistletoe haustoria among all mistletoes.

Pollination varies, as does fruit diaspore (Kirkup, 1998). Except for species of *Arceuthobium* (which have explosive fruits), African mistletoes are bird dispersed. Some species are frugivore specific, others less so. However, the spread of mistletoes is almost exclusively via birds, making control more difficult (see

Fig. 2.1. (A) *Viscum minimum* (upper centre) on *Euphorbia polygona*, cultivated at the Copenhagen Botanical Garden. Because the parasite is green in colour, it is difficult to discern from the host. (B) *Viscum cruciatum* flowers, Palestinian Territory.

Fig. 2.2. Examples of Loranthaceae flowers. (A) *Tapinanthus oleifolius* Windhoek, Namibia. (B) *Oncocalyx welwitschii*, Namutoni, Namibia.

Fig. 2.3. *Oncocalyx welwitschii* (above) parasitizing *Boscia microphylla* (Capparaceae), Northern Namibia. In the haustoria nomenclature of Teixeira-Costa *et al*. (2020), this is a mixed-type haustorial system.

Section 2.6, below). Birds are attracted to the coloured (white, red, yellow) fruits, which are highly specialized for bird distribution. The birds extract the single-seeded fruit, swallow it, and then either regurgitate it or defaecate the seed on to a host branch (Fig. 2.4). The seeds are covered with a very sticky substance known as viscin. This 'bird glue' or 'bird lime' was once used to trap birds. Unlike many parasitic plants, a host stimulant is not needed to induce germination. In fact, many seeds will germinate on non-living surfaces.

If the substrate is suitable, the mistletoe hypocotyl forms an attachment disc and penetrates the host. One of the few studies on germination is for *Tapinanthus bangwensis* (Room, 1973). There are varying patterns of growth and host dependence. The mistletoe habit has produced adaptations such as geotropically neutral hypocotyls that enable the parasite to penetrate the host at differing angles. Such adaptations result in the globose growth forms of species of *Viscum* and other mistletoes (Fig. 2.5).

Numerous studies have shown that mistletoes grow best in abundant light (Gairola *et al*., 2013; Zakaria *et al*., 2014). In one study of cultivated mistletoes, Room (1973) showed that seeds of *Tapinanthus bangwensis* produced flowering

Fig. 2.4. *Tapinanthus bangwensis*, Kadugli, Sudan. (A) Seeds deposited on twig of an unknown tree by birds wiping their beak. (B) Germinating seed.

Fig. 2.5. *Viscum cruciatum* on almond (*Prunus dulcis*), near Ramallah, Palestinian Territory. The globose shape of the shrubs is evident.

plants only 10 weeks (!) after germination. For other germination studies, see Roxburgh and Nicolson (2008).

2.3 Hosts

Hosts are diverse and no doubt many additional hosts could be added to the host range of African mistletoes. Some trees are more suitable hosts than others.

Although factors determining host preference are poorly studied, it has been suggested that mistletoes favour hosts with high nutrient levels (e.g. Ehleringer *et al.*, 1986). The nutrient levels of the hosts are expressed in the leaves of mistletoes, which make them a favoured browse for animals including elephants (Midgley and Joubert, 1991). However, the patterns in mistletoe assemblages may be determined primarily by the geographic range of host genera (Griffiths *et al.*, 2016).

A compilation of economically important hosts in Africa, based on a review of the literature, herbarium specimens and the authors' field observations, is provided in Table 2.1. Little is known of their potential host range in contrast to more intensively studied parasitic weeds of crops such as broomrapes and witchweeds. A helpful summary of host ranges is found in Mathiasen *et al.* (2008).

Unusual features of mistletoe parasitism are epi- and hyperparasitism in which one mistletoe parasitizes another. Several genera in Africa display these characteristics. Some *Viscum* species, such as *V. loranthicola*, are only parasitic on other mistletoes. This obligate association is known as epiparasitism. If the

Table 2.1. Documented crop hosts of African mistletoes attacking major crops.

Mistletoe	Crop	Reference
Viscaceae		
Viscum anceps	Citrus *Citrus* sp.	Polhill and Wiens, 1998
Viscum engleri	Peach *Prunus persica*	Polhill and Wiens, 1998
Viscum (*rotundifolium*)	Olive *Olea europaea*	Dean *et al.*, 1994
Loranthaceae		
Erianthemum dregei	Marula *Sclerocarya birrea*	Gairola *et al.*, 2013
	Pecan *Carya illinoensis*	
	Mulberry *Morus alba*	
	Teak *Tectona grandis*	
Phragmanthera capitata	Cashew *Anacardium occidentale*	Dibong *et al.*, 2008;
	Citrus *Citrus* sp.	Houénon *et al.*, 2012
	Guava *Pisidium guajava*	
	Coffee *Coffea robusta*	
	Cola *Cola nitida*	
	Cacao *Theobroma cacao*	
	Avocado *Persea americana*	
Phragmanthera incana	Shea tree *Vitellaria paradoxa*	Ogunmefun *et al.*, 2013
	Cacao *Theobroma cacao*	
	Coffee *Coffea arabica*	
	Guava *Psidium guajava*	
	Orange(?) *Citrus* sp.	
	Cola *Cola nitida*	
Tapinanthus bangwensis	Orange(?) *Citrus* sp.	Asare-Bediako *et al.*, 2013
	Avocado *Persea americana*	
	Cassava *Manihot esculenta*	
	Cacao *Theobroma cacao*	
	Guava *Psidium guajava*	Musselman, 2019
	Sweet orange *Citrus* × *sinensis*	Edagbo *et al.*, 2012

Fig. 2.6. Hyperparasitism. *Tapinanthus bangwensis* (above) parasitizing *Oncocalyx welwitschii*, Windhoek, Namibia. Note the similarity of leaves in these two mistletoes.

relationship is facultative, i.e. the mistletoe may be found on another mistletoe but it may also be found attached directly to a host tree, this relationship is termed hyperparasitism. Unlike other assemblages of parasitic angiosperms, hyperparasitism is widespread among mistletoes (Fig. 2.6).

2.4 Distribution

Mistletoe '...groweth onely upon certain Trees, and not upon many whereon these Birds do light'. This quote from Thomas Browne's *Pseudodoxia Epidemica*, first published in 1672, addresses one of the continuing bafflements of mistletoe biology. Why are some trees heavily parasitized while the same species nearby is parasite free? To understand mistletoe distribution requires knowledge of the host trees and bird behaviour. Because of their obligate relationship with birds, the architecture of the host tree can play an important role in the number of bird visitors and the resultant deposition of parasite seeds. This has been demonstrated by Roxburgh and Nicolson (2008) who found that mistletoe survival was greater in taller trees, but that germination and attachment of a mistletoe were not dependent on its placement height-wise in the tree. This clearly shows the importance of understanding the relationship between birds, mistletoes and host trees. Likewise, studying bird behaviour for both pollination and fruit dispersal is necessary (N. Flanders and L.J. Musselman, personal communication). Understanding of the mechanisms behind the spread of mistletoes is still limited, as is the knowledge of their actual distribution; this is partly because of the lack of detection of the parasite in the host when it is not in flower, due to leaf shape mimicry and the host blocking sight of the mistletoe.

Based on herbarium records, distribution maps of different mistletoe species with relevance to African agriculture are made. In Africa, *Tapinanthus bangwensis* occurs in the Guinean forest–savannah and into the Sudan savannah zones of Western Africa, with some observations south of the equator in the Democratic Republic of the Congo (DRC) and Malawi. *Tapinanthus belvisii* seems more restricted to the Guinean forest–savannah zones of Western Africa, with most records from Côte d'Ivoire and Ghana (Fig. 2.7). *Erianthemum dregei* is mainly found in eastern and southern parts of the continent, except Madagascar, but it is also recorded in Togo (Fig. 2.8). *Phragmanthera incana* is observed in southern (coastal) parts of Côte d'Ivoire, Ghana and Nigeria, as well as in Cameroon. *Phragmanthera capitata* seems to have a wider distribution, from the Guinean forest–savannah of Western Africa to the rainforest areas of the Congo Basin (Fig. 2.9). The *Viscum* species are found in Northern, Eastern and Southern Africa (Fig. 2.10). *Viscum anceps* is recorded in South Africa and Burundi, *V. cruciatum* is known only from Morocco, whereas *V. engleri* is recorded in Equatorial Guinea, Tanzania and Malawi. The most widespread *Viscum* species in Africa is *V. rotundifolium*, which is reported across Southern Africa up to DRC.

Fig. 2.7. *Tapinanthus bangwensis* and *T. belvisii* distribution within Africa (mapped distribution depends on extent of collections). Data from Global Biodiversity Information Facility: GBIF.org (accessed 31 August 2022), GBIF Occurrence Download https://doi.org/10.15468/dl.5x67tf and https://doi.org/10.15468/dl.w3evfk. Blue and red dots indicate georeferenced records from GBIF and non-grey colours indicate all countries where the species has been observed.

Fig. 2.8. *Erianthemum dregei* distribution within Africa (mapped distribution depends on extent of collections). Data from GBIF.org (accessed 31 August 2022), GBIF Occurrence Download https://doi.org/10.15468/dl.q75v6r. Black dots indicate georeferenced records from GBIF and non-grey colours indicate all countries where the species has been observed.

No mistletoe is listed on the United States Department of Agriculture list of introduced, invasive and noxious plants. This is because there are very few mistletoes that are invasive even though several important tree crops are significantly damaged by the parasites. The range of some mistletoes is expanding, not because they are invasive but because their habitat is invaded. *Viscum album* was introduced to California over a century ago and although it has spread (Mathiasen *et al.*, 2008), it is not a threat to the multi-million-dollar crops in that state. Mistletoe damage in Ghana to cacao, a native of the New World, is one of several examples of non-autochthonous hosts attacked by native mistletoes. Perhaps it is not surprising that introduced trees seem more susceptible to mistletoe damage than indigenous trees. For example, shea tree, native to Western Africa, can have a considerable parasite load but with little damage (Lamien *et al.*, 2004). In fact, there

Fig. 2.9. *Phragmanthera incana* and *P. capitata* distribution within Africa (mapped distribution depends on extent of collections). Data from GBIF.org (accessed 31 August 2022), GBIF Occurrence Download https://doi.org/10.15468/dl.ddrmnj and https://doi.org/10.15468/dl.6pm5wp. Blue and red dots indicate georeferenced records from GBIF and non-grey colours indicate all countries where the species has been observed.

was no significant difference between flowering and fruiting of parasitized and non-parasitized branches.

2.5 Host Damage

At their simplest, loranthaceous mistletoes divert resources from the host's vascular system so that parts distal to the haustorial attachment senesce and die. Other imposed damage, as expected from a plant parasite, is a shunting of resources in the host vascular system. Symptoms of mistletoe damage are summarized in Mathiasen *et al.* (2008). Typical host damage is shown in Fig. 2.11.

Fig. 2.10. *Viscum anceps*, *V. cruciatum*, *V. engleri* and *V. rotundifolium* distribution within Africa (mapped distribution depends on extent of collections). Data from GBIF.org (accessed 31 August 2022), GBIF Occurrence Download https://doi.org/10.15468/dl.c5wu8p, https://doi.org/10.15468/dl.s2sgut, https://doi.org/10.15468/dl.gbwer4 and https://doi.org/10.15468/dl.ehsbqp. Dots indicate georeferenced records from GBIF and non-grey colours indicate all countries where the species has been observed.

2.6 Control

Because mistletoe parasitism is in essence a well-tuned graft, control is difficult without damaging the host. Little information is available on the control of African mistletoes; a global review of control measures is found in Mathiasen *et al.* (2008).

In rubber plantations, cutting the mistletoe has been standard practice. However, some mistletoes produce cortical strands, strands of tissue that grow under the bark and can break through the bark to produce a new shrub. For this reason, systemic control would be ideal. This has been attempted by trunk injection of herbicides in Australia to control mistletoe in eucalyptus

Fig. 2.11. Guava (wider leaves) parasitized by *Tapinanthus bangwensis*, Abu Naama, Sudan. Dead branches of the host are distal to the point of attachment of the haustorium (arrow).

(Greenham and Brown, 1957). Control of mistletoes on shade trees (15 species in diverse genera) in California used injections combined with pruning (Michailides *et al.*, 1987). These and other studies have shown the difficulty of mistletoe control because of resprouting of the mistletoe from cortical strands or similar reproductive structures, and the impossibility of controlling deposition of additional seeds by birds. Accordingly, the smallholder farmer has few alternatives to mechanical control except excision of the parasite (e.g. Asare-Bediako *et al.*, 2013).

2.7 Use of Mistletoe

The ecological value of mistletoes is now recognized as being keystone species in some systems (see Mathiasen *et al.*, 2008). Economic value in terms of products of African mistletoes includes medicine and wood roses (Dzerefos *et al.*, 1999).

2.7.1 Medicine

Like other places in the world, mistletoes in Africa have many uses in indigenous medicines. Adesina *et al.* (2013) review traditional uses of mistletoe, including ethnopharmacology. In some places mistletoes are grown as a sort of crop (Ogunmefun *et al.*, 2013) to produce medical products.

2.7.2 Wood roses

The host–mistletoe connection can be thought of as a placenta where the tissues of the two organisms come into intimate contact. In some mistletoes this

Fig. 2.12. Tree rose of unknown host resulting from parasitism from undetermined mistletoe. Purchased in Gauteng Province, South Africa.

haustorium grows to become quite large through specialized cambia, well-coordinated with host tissues. When the parasite dies, the mistletoe tissue disintegrates and leaves the host portion of the haustorium. These structures are called wood roses and are sold as curios (Dzerefos *et al.*, 1999; Fig. 2.12).

References

Adesina, S., Illoh, H., Johnny, I. and Jacobs, I. (2013) African mistletoes (Loranthaceae); ethnopharmacology, chemistry and medicinal values: an update. *African Journal of Traditional, Complementary and Alternative Medicines* 10, 161–170.

Asare-Bediako, E., Addo-Quaye, A., Tetteh, A., Buah, J., Van Der Puije, G. *et al.* (2013) Prevalence of mistletoe on citrus trees in the Abura-Asebu-Kwamankese District of the Central Region of Ghana. *International Journal of Scientific and Technology Research* 2, 122–127.

Browne, T. (1672) *Pseudodoxia Epidemica: or, Enquiries into Commonly Presumed Truths*. Benediction Classics, Garsington, UK (2009 reprint).

Calder, D. and Bernhardt, P. (eds) (1984) *The Biology of Mistletoes*. Academic Press, Cambridge, Massachusetts.

Calvin, C. and Wilson, C. (1998) Comparative morphology of haustoria within African Loranthaceae. In: Polhill, R. and Wiens, D. *Mistletoes of Africa*. Royal Botanic Gardens, Kew, UK, pp. 17–36.

Dean, W., Midgley, J. and Stock, W. (1994) The distribution of mistletoes in South Africa: patterns of species richness and host choice. *Journal of Biogeography* 21, 503–510.

Dibong, D., Din, N., Priso R., Desiré, T., Fankem H. *et al.* (2008) Parasitism of host trees by the Loranthaceae in the region of Douala (Cameroon). *African Journal of Environmental Science and Technology* 2, 371–378.

Dzerefos, C., Shackleton, C. and Witkowski, E. (1999) Sustainable utilization of woodrose-producing mistletoes (Loranthaceae) in South Africa. *Economic Botany* 53, 439–477.

Edagbo, D., Ajiboye, T., Temitope, T., Ighere, D., Alwonle, A. *et al.* (2012) The influence of African mistletoe (*Tapinanthus bangwensis*) on the conservation status of *Citrus sinensis* in Moor Plantation Area of Ibadan, Nigeria. *International Journal of Current Research* 4, 484–487.

Ehleringer, J., Ullmann, I., Lange, O., Farquhar, G., Cowan, I. *et al.* (1986) Mistletoes: a hypothesis concerning morphological and chemical avoidance of herbivory. *Oecologia* 70, 234–237.

Gairola, S., Bhatt, A., Govender, T., Faijnath, H., Procheș, Ș. *et al.* (2013) Incidence and intensity of tree infestation by the mistletoe *Erianthemum dregei* (Eckl. & Zeyh.) V. Tieghem in Durban, South Africa. *Urban Forestry & Urban Greening* 12, 315–322.

Glatzel, G., Richter, H., Devkota, M., Amico, G., Lee, S. *et al.* (2017) Foliar habit in mistletoe–host associations. *Botany* 95, 219–239.

Greenham, C. and Brown, A. (1957) The control of mistletoe by trunk injection. *Journal of the Australian Institute of Agricultural Science* 23, 308–318.

Griffiths, M., Ruiz, N. and Ward, D. (2016) Mistletoe species richness patterns are influenced more by host geographic range than nitrogen content. *African Journal of Ecology* 55, 101–110.

Heide-Jørgensen, H. (2008) *Parasitic Flowering Plants*. Brill, Leiden, The Netherlands.

Houénon, J., Yedomonhan, H., Adomou, A., Tossou, M., Madjidouet, M. *et al.* (2012) Les Loranthaceae des zones guinéenne et soudano-guinéenne au Bénin et leurs hôtes. *International Journal of Biological and Chemical Sciences* 6, 1669–1686.

Kirkup, D. (1998) Pollination mechanisms in African Loranthaceae. In: Polhill, R. and Wiens, D. *Mistletoes of Africa.* Royal Botanic Gardens, Kew, UK, pp. 37–60.

Kuijt, J. (1969) *Parasitic Flowering Plants*. University of California Press, Berkeley, California.

Lamien, N., Boussim, J., Nygard, R., Ouédraogo, J., Odén, P., *et al.* (2004) Mistletoe impact on shea tree (*Vitellaria paradoxa* C. F. Gaertn.) flowering and fruiting behavior in savanna area from Burkina Faso. *Environmental and Experimental Biology* 5, 142–148.

Mathiasen, R., Shaw, D., Nickrent, D. and Watson, D. (2008) Mistletoes. Pathology, systematics, ecology, and management. *Plant Disease* 92, 988–1006.

Michailides, T., Ogawa, J., Parmeter, J. Jr and Yoshimine, S. (1987) Survey for and chemical control of leafy mistletoe (*Phoradendron tomentosum* subsp. *microphyllum*) on shade trees in Davis, California. *Plant Disease* 71, 533–536.

Midgley, J. and Joubert, D. (1991) Mistletoes, their host plants and the effects of browsing by large mammals in Addo Elephant National Park. *Koedoe* 34, 149–152.

Musselman, L.J. (2019) *Parasitic Weeds in African Agriculture*. ECHO Technical Note 94, 12 pp.

Nickrent, D. (2020) Parasitic angiosperms: how often and how many? *Taxon* 69, 5–27.

Ogunmefun, O.T., Fasola, T.R., Saba, A.B. and Oridupa, O.A. (2013) The ethnobotanical, phytochemical and mineral analyses of *Phragmanthera incana* (Klotzsch), a species of mistletoe growing on three plant hosts in South-Western Nigeria. *International Journal of Biomedical Science* 9, 33–40.

Parker, C. and Riches, C. (1993) *Parasitic Plants of the World: Biology and Control.* CAB International, Wallingford, UK.

Polhill, R. and Wiens, D. (1998) *Mistletoes of Africa*. Royal Botanic Gardens, Kew, UK.

Room, P. (1973) Ecology of the mistletoe *Tapinanthus bangwensis* growing on cocoa in Ghana. *Journal of Ecology* 61, 729–742.

Roxburgh, L. and Nicolson, S. (2008) Differential dispersal and survival of an African mistletoe: does host size matter? *Plant Ecology* 195, 21–31.

Teixeira-Costa, L., Ocampo, G. and Ceccantina, G. (2020) Morphogenesis and evolution of mistletoes' haustoria. In: Demarco, D. (ed.) *Plant Ontogeny: Studies, Analyses and Evolutionary Implications*. Nova Science, Hauppauge, New York, pp. 108–157.

Visser, J. (1981) *South African Parasitic Flowering Plants*. Juta, Cape Town.

Zakaria, R., Addo-Fordjour, P., Mansor, A. and Fadzly, N. (2014) Mistletoe abundance, distribution and associations with trees along roadsides in Penang, Malaysia. *Tropical Ecology* 55, 255–262.

3 Love Vine

Abstract

There are two representatives of the genus *Cassytha* in Africa. The only one that is a crop parasite is the pan-tropical *Cassytha filiformis* with the common name of love vine. It is often misidentified because of its remarkable similarity to another stem parasite we are considering: species of *Cuscuta*, dodder. Although *Ca. filiformis* is widespread on the continent, especially in coastal regions, it has received less attention than other groups of parasites.

3.1 Introduction

Species of *Cassytha* are climbing, dextrorsely twining vines often forming dense masses of tangled, shallowly grooved stems (Fig. 3.1). Leaves are scale-like, about 2 mm long and often lacking on mature stems. Host stems are parasitized by peg-like haustoria that penetrate the host and form a morphological and physiological bridge between host and parasite. The flowers are tiny (~5 mm wide), white, and produce a white or red fleshy single-seeded translucent fruit.

Stems are green and photosynthetic, with the chlorophyll sometimes masked by orange pigment (De La Harpe *et al.*, 1981). Although photosynthetic, the plant cannot grow without parasitizing a host. It is regularly mistaken for a species of dodder (*Cuscuta*) from which it can readily be distinguished by the presence of short hairs (evident with a 10× hand lens).

3.2 Taxonomy

The taxonomy of the genus in Africa is unsettled. The genus, part of the Lauraceae (laurel) family, reaches its greatest diversity in Australia (Weber, 1981).

 Parasitic Plants in African Agriculture (L.J. Musselman and J. Rodenburg)
DOI: 10.1079/9781789247657.0003

Fig. 3.1. *Cassytha filiformis*. (A) Twisted masses of stems wrapping around themselves. (B) Festooning an unidentified tree along the Okavango River, Namibia. (C) Stem tip showing hairs. The orange colour varies from plant to plant and may mark the parasite's lessened dependence upon its own photosynthesis. (D) Scanning electron micrograph of a stem of *Ca. filiformis* showing the hairs and rows of stomata.

There are perhaps three species in Africa (Visser, 1981; Diniz, 1996): *Cassytha ciliolata* Nees, *Ca. filiformis* L. and *Ca. pondoensis* Engl. *Cassytha ciliolata* is restricted to South Africa and has red fruits. The third species, *Ca. pondoensis*, could be a segregate of *Ca. filiformis*. Species are separated on technical characteristics of the tepals but there are intermediates. Except for *Ca. ciliolata*, the African species resemble one another and behave similarly and are therefore treated together under *Ca. filiformis*.

3.3 Biology

Flowers and fruits can be produced in different seasons. The floral biology of the parasite is poorly studied and pollinators are unknown. Tiny white flowers develop into white fleshy fruits (Fig. 3.2).

Usually only one of the flowers on the spike will produce a fruit. Fruits are globular and about 1 cm in diameter. *Cassytha filiformis* fruits surround a single black seed. Technically, the fruit is an accessory fruit as the fleshy portion arises from the receptacle of the flower, not the ovary wall (Fig. 3.3).

Fig. 3.2. Love vine flowers are small and borne on short spikes. Note the green stem; the amount of chlorophyll varies among plants.

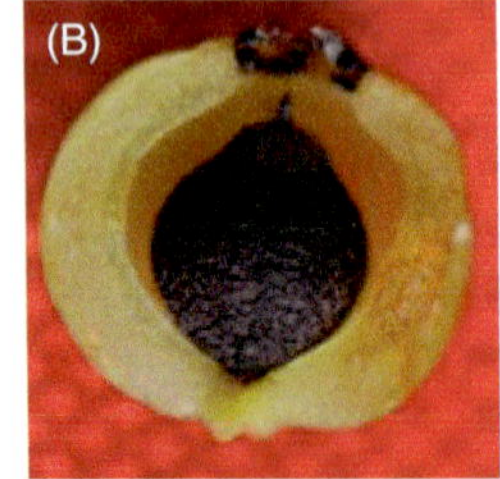

Fig. 3.3. Love vine fruits. (A) The middle fruit is not quite ripe and retains some chlorophyll. (B) Section of fruit showing the fleshy covering derived from the receptacle of the flower and the hard, black seed.

The wall of the seed (technically the fruit) is indurated. The fleshy structure of the fruit suggests bird distribution. Mahadevan and Jayasuriya (2013) have studied dormancy in *Ca. filiformis*, showing the role of scarification in germination. The need for scarification could explain the presence of this parasite along beaches and watercourses where wave action would be an agent of scarification, as with *Cuscuta*. A similar phenomenon occurs when soils are disturbed for agriculture.

Germination is epigeal and cryptocotylar; the cotyledons remain in the seedcoat, as shown for the Australia species *Ca. glabella* R. Br. (Clifford, 1999). The process appears to be the same in *Ca. filiformis* (L.J. Musselman, personal observation) (Fig. 3.4). The seedlings are green and produce what appear to be adventitious roots at the tip of the hypocotyl. To our knowledge, there are no studies of how seedlings make the initial parasitic contact and haustorial connection with their hosts.

Studies on the physiology and resultant host damage by *Ca. filiformis* and other species in the genus are sparse. Working with the Australian species *Ca. pubescens* R. Br., Shen *et al.* (2006) and Prider *et al.* (2009) showed that the host, *Cytisus scoparius* (L.) Link, was more susceptible to photodamage than uninfected plants. Burch (1992) also describes host damage. The parasitic efficiency may be aided by a very strong transpirational pull, evidenced by the abundant stomata (Fig. 3.1D).

3.4 Distribution

Cassytha filiformis is tropical and subtropical and widespread throughout Africa, although most frequent in coastal areas and along rivers (Musselman, 2019). It occurs from sub-Sahelian Africa southward to South Africa (Fig. 3.5).

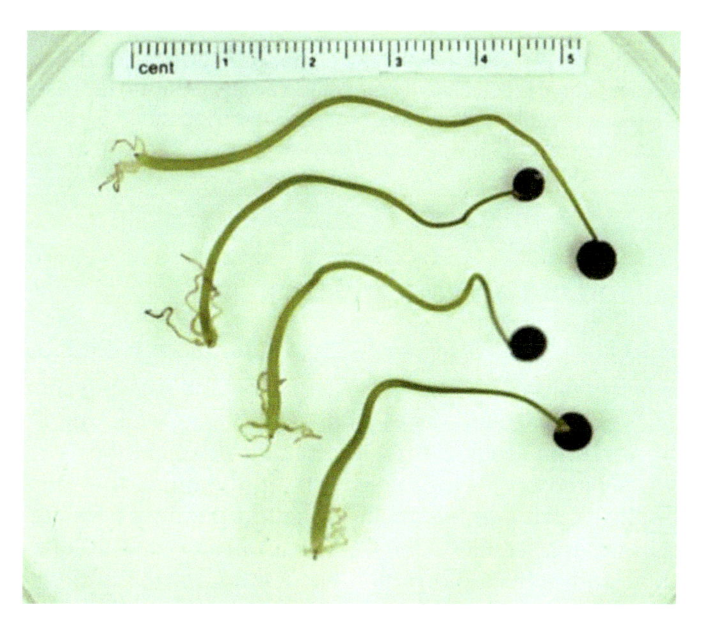

Fig. 3.4. Seedlings of *Cassytha filiformis*. The cotyledons remain in the black, globular seedcoat. At the opposite end of the seedlings, the swollen tip of the hypocotyl is evident. Here the torturous adventitious roots arise.

Fig. 3.5. *Cassytha filiformis* distribution within Africa (mapped distribution depends on extent of collections). Data from Global Biodiversity Information Facility: GBIF.org (accessed 31 August 2022), GBIF Occurrence Download https://doi.org/10.15468/dl.pp4abm. Black dots indicate georeferenced records from GBIF and non-grey colours indicate all countries where the species has been observed.

3.5 Comparison of *Cassytha* and *Cuscuta*

One of the most striking examples of parallel evolution or homoplasy in angiosperms is this pair of parasites, *Cassytha* spp. and *Cuscuta* spp. Both obligate hemiparasites are leafless vines, similar in colour, attacking a wide diversity of hosts, and with seeds that require scarification. Yet they are unrelated and widely separated on phylogenetic trees. Furthermore, they are unique within their respective families, so much so that taxonomists have constructed separate monogeneric families – the Cassythaceae and the Cuscutaceae. However, molecular studies have shown that the genus *Cassytha* is clearly nested in the Lauraceae, the laurel family consisting of trees and shrubs except for *Cassytha*. *Cuscuta* clearly belongs in the Convolvulaceae, the morning glory family, although it is the only parasitic genus of that family (phylogeny summarized in

Nickrent, 2020). Again, despite the evolutionary distance between these two parasitic vines, they are repeatedly confused in the literature and herbarium collections. Because their agronomic impact, host selection and control are so different, it is important to point out the distinctions as well as using non-confusing common names (Table 3.1). For discussion of *Cuscuta* spp., see Chapter 4, this volume. See also Haynes *et al.* (1996) for comparisons.

3.6 Host, Damage and Control

The host range of *Cassytha filiformis* is very broad and we have little idea of host selection and preference. Unlike *Cuscuta* species, *Ca. filiformis* will parasitize monocots, but to date there is no report of damage to grains. Although there is no quantitative data, it does appear that this species – at least in Africa – prefers woody hosts. A compilation of economically important hosts, based on a review of the literature, herbarium specimens and the authors' field observations, is provided in Table 3.2.

The greatest documented damage to an African crop is to cashew in Tanzania (Buriyo *et al.*, 2016). It also occurs in Western Africa on cashew and mango (Musselman, 2019). There is one account of avocado damage in the coastal region of Tanzania. In other parts of the world, it has been reported on crops important in Africa, including citrus, coconut and cassava. Although extensive damage is limited, it is important to note that it could become a serious problem in plantations. Detailed host injury by the parasite has received

Table 3.1. Comparison between *Cassytha* spp. and *Cuscuta* spp.

Feature	*Cassytha*	*Cuscuta*
Distribution	Tropical and subtropical	Most species temperate
Indumentum	Short hairs	Glabrous
Flowers	Trimerous	Tetramerous or pentamerous
Fruit	Single-seeded, fleshy	Capsule with three or four seeds
Habit	Perennial	Annual
Odour	Pungent when crushed	No odour

Table 3.2. Economically important hosts of *Cassytha filiformis*.

Host-crop common name	Host-crop scientific name	Reference
Coconut	*Cocos nucifera*	Waterhouse, 1997
Avocado	*Persea americana*	Schroeder, 1978
Mango	*Mangifera indica*	Musselman, 2019 (Fig. 3.6A)
Cashew	*Anacardium occidentale*	Buriyo *et al.*, 2016 (Fig. 3.6B)
Citrus	*Citrus* spp.	Beattie *et al.*, 2008
Cassava	*Manihot esculenta*	Herbarium specimen, JSTOR Plants

Fig. 3.6. Infestations of *Cassytha filiformis*. (A) *Ca. filiformis* on mango in Senegal. (B) *Ca. filiformis* parasitizing cashew, Guinea Conakry.

little study. Prider *et al.* (2009) reported damage by the Australian species *Cassytha pubescens* to the introduced *Cytisus scoparius* (L.) Link. Most of the serious harm is caused by smothering the trees with dense, tangled stems that impede light, leading to death of the branches.

Control is by physical removal of the parasite or, in extreme cases, by pruning the infected branches (Kidunda *et al.*, 2017). It appears the Tanzanian cashew infestation is of recent occurrence, suggesting the need for sanitation measures that prohibit the introduction of infected stock as well as surveying planned orchards for the presence of the parasite.

Although best known for its parasitic behaviour, *Ca. filiformis* and other species in the genus are used in traditional medicine as well as pharmacological surveys (Nelson, 2008).

References

Beattie, G.A.C., Holford, P., Mabberley, D.J., Haigh, A.M. and Broadbent, P. (2008) On the origins of citrus, huanglongbing, *Diaphorina citri* and *Trioza erytreae*. In: Gottwald, T.R. and Graham, J.H. (eds) *Proceedings of the 2008 International Research Conference on Huanglongbing.* Plant Management Network, Orlando, Florida, pp. 23–56. Available at: www.plantmanagementnetwork.org/proceedings/irchlb/2008/presentations/IRCHLB.K.2.pdf (accessed 5 May 2023).

Burch, J.N. (1992) *Cassytha filiformis* and limits to growth and reproduction of *Schinus terebinthifolius* in southern Florida. *Florida Scientist* 55, 28–34.

Buriyo, A.S., Kasuga, L., Moshi, H.N. and Nene, W.A. (2016) Ecological distribution and abundance of the parasitic weed, *Cassytha filiformis* L. (Lauraceae) in major cashew, *Anacardium occidentale* L., growing regions in Tanzania. *International Journal of Basic and Applied Sciences* 5, 109–116.

Clifford, H.T. (1999) The seedling of *Cassytha glabella* R. Br. *Austrobaileya* 5, 345–347.

De La Harpe, A.C., Visser, J.H. and Grobbelaar, N. (1981) Photosynthetic characteristics of some South African parasitic flowering plants. *Zeitschrift für Pflanzenphysiologie* 103, 265–275.

Diniz, M.A. (1996) Lauraceae of the Flora Zambesiaca area. *Kirkia* 16, 55–68.
Haynes, A.R., Coile, N.C. and Schubert, T.S. (1996) Comparison of two parasitic vines: dodder (*Cuscuta*) and woe vine (*Cassytha*). *Botany Circular Number 30.* Division of Plant Industry, Florida Department of Agriculture and Consumer Services, Tallahassee, Florida.
Kidunda, B.R., Kasuga, L.J. and Alex, G. (2017) Assessing the existence spread and control strategies of parasitic weed (*Cassytha filiformis*) on cashew trees in Tanzania. *Journal of Advanced Agricultural Technologies* 4, 285–289.
Mahadevan, N. and Jayasuriya, K.M.G.G. (2013) Water-impermeable fruits for the parasitic angiosperm *Cassytha filiformis* (Lauraceae): confirmation of physical dormancy in Magnoliidae and evolutionary considerations. *Australian Journal of Botany* 61, 322–329.
Musselman, L.J. (2019) *Parasitic Weeds in African Agriculture*. ECHO Technical Note 94, 12 pp.
Nelson, S.C. (2008) *Cassytha filiformis*. Plant Disease PD-42 Cooperative Extension Service, College of Tropical Agriculture and Human Resources, University of Hawai'i at Mānoa, 10 pp.
Nickrent, D.L. (2020) Parasitic angiosperms: how often and how many? *Taxon* 69, 5–27.
Prider, J., Watling, J. and Facelli, J.M. (2009) Impacts of a native parasitic plant on an introduced and a native host species: implications for the control of an invasive weed. *Annals of Botany* 103, 107–115.
Schroeder, C.A. (1978) An unusual case of parasitism in avocado. *California Avocado Society Yearbook* 62, 126–130.
Shen, H., Ye, W., Hong, L., Huang, H., Wang, Z. *et al.* (2006) Progress in parasitic plant biology: host selection and nutrient transfer. *Plant Biology* 8, 175–185.
Visser, J. (1981) *South African Parasitic Flowering Plants.* Juta, Cape Town.
Waterhouse, D.F. (1997) *The Major Invertebrate Pests and Weeds of Agriculture and Plantation Forestry in the Southern and Western Pacific*. Monograph Number 44, Australian Centre for International Agricultural Research, Bruce, Australia, 99 pp.
Weber, J.Z. (1981) A taxonomic revision of *Cassytha* (Lauraceae) in Australia. *Journal of the Adelaide Botanical Garden* 3, 187–262.

4 Dodder

Abstract
Dodders, species of the genus *Cuscuta,* are one of the largest groups of parasitic angiosperms in Africa, with 11 different species known as weeds, and are readily noticeable by their bright yellowish-orange or reddish thread-like stems. These stems often form large, entwined masses, appearing like a tangled heap of spaghetti, that can smother the host. Host selection and host specificity vary immensely among dodder species; some, like the broadly distributed *C. campestris,* have a wide diversity of hosts of different families and may even attach to a range of hosts at one time, whereas others are usually found only on a single host. Several species can cause serious economic loss. Some are introduced parasites expanding their range but there are also numerous native species. Because of considerable confusion with species determinations of these pests, we discuss features useful for identification.

4.1 Introduction

Species of *Cuscuta*, known as dodders in English, are annual (rarely perennating in host tissue), climbing, twining vines, often forming dense masses of tangled yellow, orange or reddish stems that parasitize host stems. They are among the most glabrous (lacking hairs) of any plant. Leaves are scale-like, about 2 mm long and occur only at branches. Dodders are stem parasites and penetrate their hosts by peg-like haustoria forming a morphological and physiological bridge linking host and parasite. Rows of haustoria often form a chain-like series. Although they are obligate parasites that cannot complete their life cycle without a host, most species of dodder have some chlorophyll, which is often most apparent in seedlings and developing fruits. The white or pink flowers are small and in arrangements of four or five. The fruit is a papery capsule with one to four seeds.

Cuscuta is the only parasitic genus in the Convolvulaceae (the morning glory family). Because of its parasitic habit, it is sometimes treated by some authors as

DOI: 10.1079/9781789247657.0004

a separate family, the Cuscutaceae. However, molecular phylogenetic research supports *Cuscuta* as a member of the Convolvulaceae (Costea *et al.*, 2015).

Helpful reviews of the biology and control of *Cuscuta* spp. are found in Kuijt (1969), Parker and Riches (1993), Dawson *et al.* (1994), Costea and Tardif (2005), Lanini and Kogan (2005), Heide-Jørgensen (2008) and Costea *et al.* (2015). There has been a great deal of attention paid to the genus in recent years, as shown by a simple Google search yielding over 1 million results. Not a small portion of these results are due to the widespread use of dodder in traditional medicines (several reviews online; see also Costea and Tardif, 2005).

In this chapter, we provide an overview of African *Cuscuta* spp., their biology, agronomic importance and control.

4.2 Taxonomy and Identification

A review of the taxonomy of this group is important because of the frequent misidentifications appearing in the literature and requests to us for species determinations. A chief culprit of mistaken identity is *Cuscuta campestris*. Because of its widespread distribution, there is a tendency to assume a new infestation of dodder outside North America, where it is native, to be this species, especially if the plants are yellow-orange. This error is even more likely for those species that are superficially similar, such as *C. australis* and *C. chinensis*. Although minor morphological differences exist between *C. campestris* and *C. pentagona* (according to Yuncker, 1932), these species are often treated as one (e.g. Beliz, 1986) under the name *C. pentagona* Engelm.

Like other groups with taxonomic and identification complexity, *Cuscuta* has benefited from numerous recent phylogenies. We are using the comprehensive systematic treatment of subgeneric classification by Costea *et al.* (2015) who recognize four subgenera: *Monogynella*, *Grammica*, *Cuscuta* and *Pachystigma*. African species in the subgenus *Pachystigma* are restricted to South Africa; none is known to be of agronomic importance, and they are not discussed further in this volume. The subgenus *Monogynella* has only two African representatives: *C. monogyna*, which is a documented problem in some tree crops and occurs only in the Mahgreb countries and Egypt; and *C. cassytoides* in South Africa, where it has not been a problem. Like other members of the subgenus, these two species are robust, high-climbing vines with distinct flowers, but they are the only *Cuscuta* species to have seeds with smooth surfaces. The largest group of African dodders is in the subgenus *Grammica*. Species of documented agronomic importance in this group include *Cuscuta australis*, *C. campestris*/*C. pentagona*, *C. chinensis*, *C. hyalina*, *C. kilimanjari* and *C. suaveolens*. The subgenus *Cuscuta* includes *C. epilinum*, *C. epithymum*, *C. palaestina*, *C. pedicellata* and *C. planiflora*. These are much more delicate plants, more restricted in their agronomic distribution and host damage than species in *Grammica*. In general, species in the subgenus *Cuscuta* favour cooler climates, such as in tropical highlands, and they also flower and fruit earlier.

Distinguishing the subgenus of a dodder is an easy first step to determining the species, especially if flowers are present. Flower structure, especially

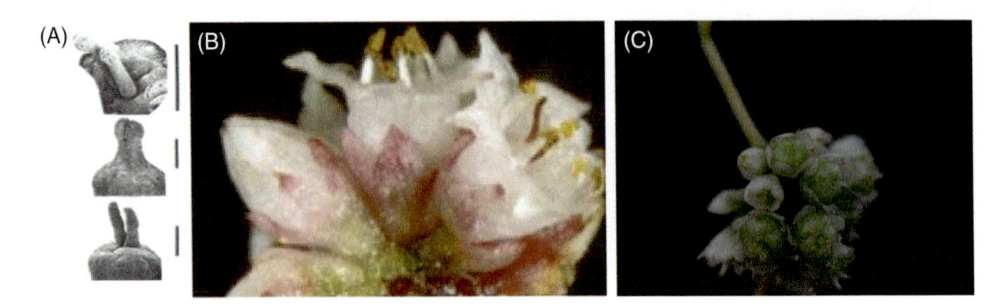

Fig. 4.1. Dodder. (A) Scanning electron micrographs of styles at 1.0 mm scale: top: *Cuscuta campestris* with capitate stigmas characteristic of subgenus *Grammica*; middle: fused styles of *C. monogyna* characteristic of subgenus *Monogynella*; bottom: linear stigmas in *C. epithymum* characteristic of subgenus *Cuscuta*. (B) Flowers of *C. epithymum* with red linear styles evident in lower flower, subgenus *Cuscuta*. (C) Flowers of *C. australis* with green ovaries and two capitate stigmas, subgenus *Grammica*. (Fig. 4.1A from Dawson *et al.*, 1994, by permission of Review of Weed Science.)

style morphology, differs among subgenera and is a dependable way to determine subgenera and facilitate species determination. Styles are fused in *Monogynella*, separate and linear in *Cuscuta*, and separate but with capitate stigmas in *Grammica* (Fig. 4.1A).

4.3 Biology

Little is known of the pollination biology of many dodder species, and it is likely that those with tiny flowers, such as *Cuscuta planiflora*, are autogamous (self-fertilizing). Flowers are produced in large numbers and are essential for determination of species (Fig. 4.1B). After fertilization, the capsules develop, often displaying conspicuous chlorophyll (Fig. 4.1C). When mature, capsules are brown and papery (Fig. 4.2A). Species can have circumscissile dehiscence (capsule opens at base), irregular dehiscence (capsule does not open along a predetermined line) or be indehiscent. Capsule and seed size varies among species. Most capsules have four seeds. Production is prodigious; one large plant can produce thousands of seeds. Seed characteristics can be used for identification; see for example Knepper *et al.* (1990) and Olszewski *et al.* (2020).

Seeds of most weedy species can remain viable for up to ten years in dry storage (L.J. Musselman, personal observation). The seeds are refractory and require scarification to germinate. Without scarification, germination in *Cuscuta campestris*, for example, is only about 10% (Dawson *et al.*, 1994). Under natural conditions, some dodders may favour naturally disturbed sites such as stream margins where seeds are scarified by water flow (*C. gronovii*), grazing areas (*C. planiflora*) or fire sites (several species in various ecosystems, especially in Mediterranean maquis). Soil disturbance by ploughing can also be a mechanism of scarifying.

Dodders parasitize stems, so the mature seeds are often produced next to the fruits of the host and are therefore harvested with them. Seed cleaning is especially difficult when there is similarity in size of the dodder seed and the host seed – as occurs, for example, in *Medicago sativa* (alfalfa, lucerne) parasitized

Fig. 4.2. *Cuscuta campestris*. (A) Mature capsules. (B) Dodder seeds with seeds of lucerne (*Medicago sativa*); lucerne seed is similar in size and shape to dodder seed but lighter in colour and with a smooth surface. (C) Scanning electron micrograph of dodder seed showing alveolate (with minute pits in a honeycomb-like pattern) seed surface. (D) Lucerne seed contaminated with abundant dodder seed at a market in Khartoum, Sudan.

by *Cuscuta campestris*, especially in fields where the crop is grown for seed production (Fig. 4.2B). Cleaning contaminated lucerne seeds can be done by using a felt roller. Lucerne seed has a smooth seedcoat so the rough-surfaced dodder seed adheres to the felt and the lucerne seed passes through, leaving the parasite seeds on the felt. Most dodder seeds are pitted when dry (Fig. 4.2C) but after imbibition of water become papillose (Olszewski *et al*., 2020).

As with most weeds, the spread of dodder is largely anthropogenic. Dodders are spread almost entirely by seeds. However, under some conditions, portions of the parasite along with their accompanying hosts are cut and then infest new hosts, as can happen with mowing along roadsides (Chak *et al*., 2010). Segments of infected host are carried by the cutting blades. Reports of distribution by birds need confirmation. The establishment of some dodder species along streams suggests that the capsules can be carried by water.

Because of the delayed germination of dodder seeds, scarification is necessary for research projects. Sulfuric acid is one method to achieve this (Dawson *et al*., 1994). Ghantous and Sandler (2012) report a technique using a handheld rotary tool.

Except for *C. monogyna*, germination of the dodders considered here is cryptocotylar (cotyledons are retained within the seedcoat) and the epicotyl emerges and then searches for a host. There are no roots in the usual sense of the term and for a short time the seedling is autotrophic. Dodder branches exhibit thigmotropism (responding to touch), which induces coiling (Fig. 4.3) around a potential host (although inanimate objects can also induce coiling).

Stems locate hosts through nastic movements left to right searching for a host. This is known as circumnutation, a widespread phenomenon in plants

(Stolarz, 2009). Once attachment is made, the parasite will grow rapidly and can blanket the host in a short time (Fig. 4.4). Damage can be so severe as to result in total crop loss.

The growth pattern of most species of the subgenus *Grammica* is distinct from that of species in the subgenus *Cuscuta*. Most *Grammica* have dimorphic branching with two kinds of stems (sometimes referred to as tendrils): searching stems and attaching stems (Fig. 4.3C). The searching stem extends the reach of the parasite, and as it does it forms branches that coil counterclockwise. The attaching stem invades a host, and the searching stem continues growing but does not produce haustoria. Subgenus *Cuscuta* species, on the other hand, do not have this feature and the searching stem both searches and attaches so that any stem can produce haustoria. Noting this pattern is a quick way to determine the subgenus. Dodders are the only plants known to forage for their nutrition (Kelly, 1992).

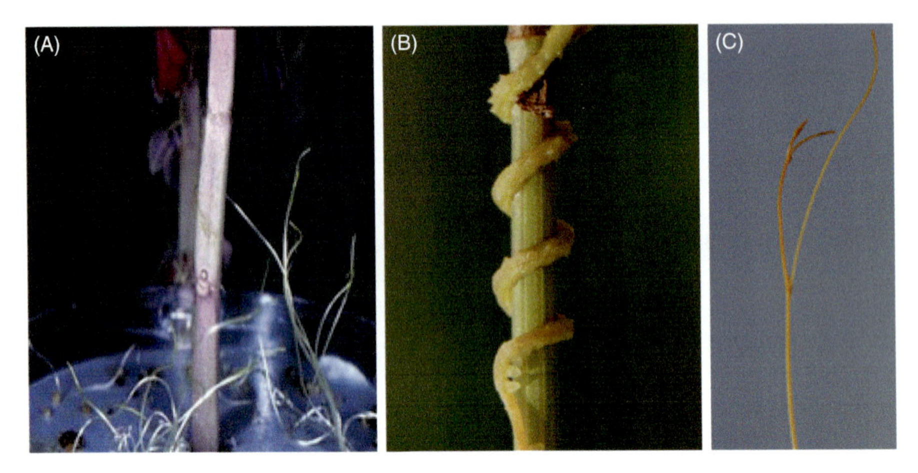

Fig. 4.3. *Cuscuta campestris*. (A) Germination and initial attachment to host, *Plectranthrus scutellarioides*. Note the chlorophyll in the seedlings. (B) Twining and attaching to *Polygonum* sp. (C) Dimorphic stems – searching and attaching stems. The branch on the left side will find and attach to a host while the one on the right side will continue growing and produce another attachment branch.

Fig. 4.4. Mature populations of *Cuscuta campestris.* (A) Smothering cumin, *Cuminum cyminum*, Aleppo, Syria. Almost all the crop is destroyed. (B) Damaging onions, *Allium cepa*, Chios, Greece. This is one of the few monocots this dodder will parasitize.

4.4 Distribution

Dodders are found throughout Africa (Fig. 4.5) in a diversity of ecosystems but generally prefer open, sunny areas. They are seldom found in forests, although some, such as *C. kilimanjari*, occur at forest margins. The most widespread

Fig. 4.5. *Cuscuta* spp. distribution within Africa (mapped distribution depends on extent of collections). Data from Global Biodiversity Information Facility: GBIF.org (5, 6 and 8 September 2022), GBIF Occurrence Downloads:
(A) *Cuscuta campestris/C. pentagona**: https://doi.org/10.15468/dl.zskfxd
(B) *Cuscuta pedicellata*, *C. planiflora*, *C. epithymum*, *C. monogyna* and *C. epilinum*: https://doi.org/10.15468/dl.xadmd4, https://doi.org/10.15468/dl.ebukj6, https://doi.org/10.15468/dl.c4shyb, https://doi.org/10.15468/dl.tvsg2c, https://doi.org/10.15468/dl.eyve9m
(C) *Cuscuta australis, C. chinensis, C. hyalina, C. kilimanjari* and *C. suaveolens*: https://doi.org/10.15468/dl.628gbu, https://doi.org/10.15468/dl.6ce7ak, https://doi.org/10.15468/dl.5sqjxz, https://doi.org/10.15468/dl.z4sqqd, https://doi.org/10.15468/dl.wzvewn
Black or coloured dots indicate georeferenced records from GBIF and non-grey colours indicate all countries where the species has been observed.
* For purposes of geographic distribution, records of *C. pentagona* are assumed to be equivalent to *C. campestris*; the GBIF database groups them together.

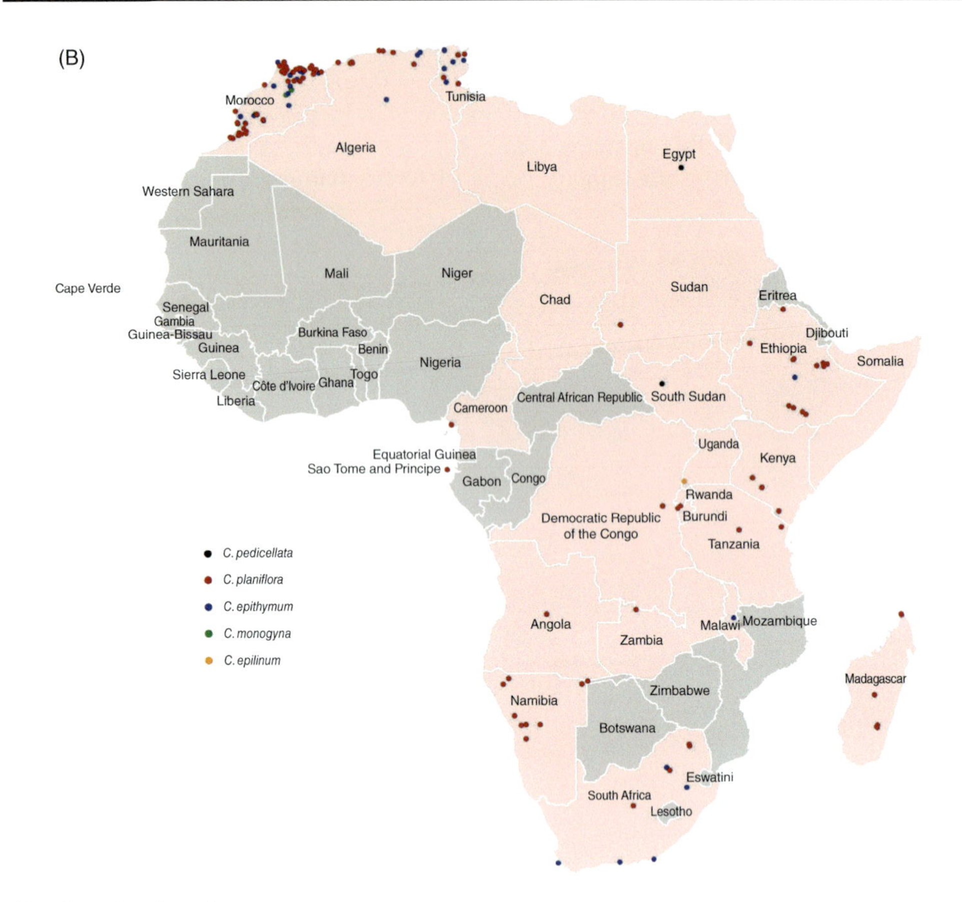

Fig. 4.5. Continued.

African dodders are *C. planiflora* (recorded in 21 countries) and *C. campestris/ C. pentagona* (18 countries). The latter has been introduced from the USA to all continents except Antarctica. Indeed, *C. campestris* is the most widely distributed parasitic weed in the world. *Cuscuta australis* (syn. *C. scandens*) is observed in 16 countries across the continent, including Madagascar (Fig. 4.5C). *Cuscuta epithymum* is observed in Northern and Southern Africa, as well as Ethiopia and Malawi, whereas *C. pedicellata* is restricted to Northern Africa, including Sudan (Fig. 4.5B). *Cuscuta hyalina* and *C. kilimanjari* have partially overlapping distributions (Burundi, Ethiopia, Kenya, Tanzania, Sudan, South Africa) but whereas the first is observed in some countries in Western Africa (e.g. Senegal) and Northern Africa (e.g. Egypt), the latter is found in the band across the Democratic Republic of the Congo, Rwanda, Malawi, Mozambique and Madagascar (Fig. 4.5C).

4.5 Species of Agronomic Importance and African Crop Hosts

Host selection and host specificity vary immensely among species of dodders. Some, such as *C. campestris/C. pentagona*, can survive on a wide diversity

Fig. 4.5. Continued.

of hosts of different families and even attach to a range of hosts at one time. Others, such as *C. epilinum*, a parasite of flax, are usually found only on a single host. Listing all the hosts and potential hosts of African dodders is a task beyond this study. We briefly discuss species that are agricultural pests, as recorded in the literature or from our own experience.

Damage to the host includes reduction in seed production and fruit quality, stunting and yellowing smothering, and resultant shading. Viruses and macromolecules can move from parasite to host through the haustoria (Bennett, 1940, 1944) and the parasite can even form a bridge through which viruses can move from one host plant to another (Birschwilks *et al.*, 2006).

It is important to distinguish between the introduced dodders that attack crops (e.g. *C. campestris* and *C. suaveolens*), and native species, which could invade agricultural fields from natural plant communities (e.g. *C. hyalina* and *C. pedicellata*). For this reason, familiarity with indigenous dodders is helpful.

A nagging problem of *Cuscuta* research is the accurate identification of dodders reported in the literature. During several decades of work in African agriculture, we have found numerous published accounts of dodder damage attributed to a misidentified parasite. Agricultural workers need to be aware of the different species to ensure proper treatment of weed problems. Accordingly, we have included this very brief synopsis of the taxonomy of the group to aid identification. More serious is confusing the unrelated *Cassytha filiformis* with species of *Cuscuta* because of the remarkable similarities in the two groups of parasites. See Chapter 3, this volume, on *Cassytha* spp.

4.6 Subgenera

This section deals with the representatives of the subgenera and their constituent species in Africa (see discussion of taxonomy above). Documentation of hosts and countries where the parasite occurs is taken from published literature, herbarium specimens (e.g. the GBIF database) and personal observations.

4.6.1 *Monogynella*

These species are the most robust of all dodders and have thick stems, distinct flowers (see above and Fig. 4.6A) and are generally restricted to woody hosts. *Cuscuta monogyna* is a pest in citrus fruits in the Middle East (Musselman, 1986) and occurs in Egypt and the Maghreb. It can cause serious damage to orchard trees (Fig. 4.6B).

Recent introductions of *Cuscuta reflexa* have been observed in tree crops in Kenya (Masanga *et al.*, 2021), which are not yet reflected in the GBIF database.

4.6.2 *Cuscuta*

There are five or so African species, only three of which have become notable agricultural problems. Overall, species in this group are more delicate than

Fig. 4.6. *Cuscuta monogyna*, Wadi al Ghor, Jordan. (A) Flowering plants on *Zizyphus spinichristi*. (B) Dodder has killed the lemon tree (*Citrus limon*) and as a result the parasite has also died.

other dodders, with thread-like stems that often have a reddish tint and very small flowers, and they are restricted to herbs and shrubs.

Cuscuta epilinum is one of the most host-specific species in the genus, being restricted almost entirely to flax, *Linum usitatissimum* (Muthaiyan, 2009). It has been reported from Egypt, Ethiopia and Morocco. It is a potential problem wherever flax is grown. *Cuscuta epithymum* attacks forage crops (Muthaiyan, 2009), is native to Egypt and the Maghreb, and apparently has been introduced to South Africa where it is not presently a problem. *Cuscuta palaestina* attacks lentil, *Lens culinaris* (Fig. 4.7). This is a species of the Eastern Mediterranean with extensions into Egypt and Libya, although no herbarium records were found in the GBIF database. It has the potential of spreading in contaminated seed (Fig. 4.2D). Like most of the species in this group, the seeds of the parasite are considerably smaller than those of the host.

Cuscuta pedicellata attacks crops such as arugula, *Eruca sativa* (Musselman, 1986; Fig. 4.8); lentil, *Lens culinaris* (Musselman *et al.*, 1989); and common vetch, *Vicia sativa* (Musselman *et al.*, 1989). Arugula has seeds similar in size to that of *C. pedicellata*. In the market in Khartoum, arugula sold as a fresh vegetable often has dodder attached with mature seed capsules. This can no doubt lead to its spread. Lentil can also be heavily infested with *C. pedicellata* (Fig. 4.7). It occurs in Egypt, Sudan and South Sudan and because of global warming could become a problem at higher elevations.

Cuscuta planiflora (Fig. 4.9) is known to parasitize forage legumes such as lucerne, *Medicago sativa* (Dawson *et al.*, 1994), and clovers where it causes serious damage. This species has the smallest flowers of any of the African dodders at just a few millimetres wide. It is a dodder of cool seasons. It is widespread from Northern to Southern Africa but mostly observed as a weed in Egypt and Ethiopia.

Fig. 4.7. *Cuscuta palaestina* parasitizing lentil, Tel Hadya, Syria. Note the yellowing lower leaves of the host.

Fig. 4.8. Field of arugula (*Eruca sativa*) infested with *C. pedicellata*, Wau, South Sudan.

Fig. 4.9. *Cuscuta planiflora*, Sulaimani Province, Iraq. (A) The red colouring is characteristic of this species, although the intensity of the colour varies considerably. (B) No stem dimorphism is evident.

4.6.3 *Grammica*

There are about seven species in this subgenus, which includes the most serious of the dodder crop pests. Vegetatively, plants can be assigned to this group by the presence of dimorphic branches (see above). The flowers are distinct with capitate stigmas (see above).

Cuscuta australis is the species most frequently mistaken for *C. campestris*. A simple field trait to separate the two is the characteristic of the tip of the corolla lobe. In *C. campestris*, the lobe is acute (pointed) with a reflexed tip. In contrast, the corolla tips of *C. australis* are obtuse (not pointed). *Cuscuta australis* is widespread, native from Senegal in the east to Ethiopia and southward to South Africa. It has a broad host range, including lucerne (Dobignard and Chatelain, 2011) and like its congener, *C. campestris*, can parasitize legume crops but usually less rigorously. The use of this species as a biological control has been suggested (Yu *et al.*, 2011) but it is imprudent to use a parasitic plant for weed control.

Cuscuta campestris (Fig. 4.10), for which the most widely used English common name is field dodder (the specific epithet *campestris* refers to fields),

Fig. 4.10. Examples of *Cuscuta campestris* infestations. (A) *Ocimum basilicum*. Note suppression of growth in heavily parasitized plants. North Carolina, USA. (B) Severely damaged *Corchorus olitorius*. Jordan Valley, Jordan. (C) Flowers showing the five-parted corolla with corolla tips reflexed. Oklahoma, USA.

is by far the most intensively studied of all species of dodder because of its wide geographic range. It is also intensively studied because it attacks a wide range of hosts, including at least 25 crop species (Lanini and Kogan, 2005). Suitable hosts include several important legume crops, such as lucerne and clover, as well as tomatoes and other vegetables. As noted earlier, it is often unjustly blamed for infestations that are indigenous *Cuscuta* species instead. It is the most damaging of any of the dodders and can occur in row crops such as *Cicer arietinum* (chickpea), *Cuminum cyminum* (cumin), *Daucus carota* (carrot), *Helianthus annuus* (sunflower), *Ocimum basilicum* (basil), *Solanum lycopersicum* (tomato) and especially forage legumes, most notably lucerne (*Medicago sativa*) and clovers (*Trifolium* spp. and others). Citrus trees (species of *Citrus*) are also attacked but with little morbidity. Many additional crops could be added.

Cuscuta chinensis, Chinese dodder, has been extensively investigated for its pharmaceutical value especially in China and has also been the subject of numerous developmental studies. It parasitizes a diversity of crops (Costea and Tardif, 2005) and is often reported as a problem in the Far East but, again, the identity of the dodder in some of these studies is questionable. At present, it is not known to be a problem in Africa where it has been reported only from Benin, Ethiopia, Sudan, Madagascar and Mauritius. A review of its ethnopharmacology documents its importance in herbal medicines (Donnapee and Li, 2014). Use of the parasite seed in herbal remedies could lead to its further dissemination. Because of the high value placed on this species for medicine, especially in China, Ren *et al.* (2020) produced a model to determine the impact of climate change on distribution that could make areas in Africa more suitable for the parasite. Their finds may be applicable to other African dodders.

Cuscuta hyalina has mainly been observed in Eastern and Southern Africa. There are few reports of its parasitism of crop plants such as arugula and lucerne (Musselman, 1986), but it poses a potential threat where native vegetation intersects with agriculture.

Cuscuta kilimanjari is certainly the most attractive dodder in Africa. It is recorded in 13 countries, most commonly in Central Africa, especially Tanzania and Rwanda where local people cultivate it by placing a mass of stems on the top of a shrub, usually bougainvillaea (Fig. 4.11A). This may be the only known case of cultivation of a dodder species as a landscape subject (Musselman, 2019). In Rwanda, *C. kilimanjari* attacks a variety of cassava (*Manihot esculenta*) grown for its leaves (Fig. 4.11B) but does not seem to occur in varieties grown for the rootstock (L.J. Musselman, personal observation).

Cuscuta suaveolens is native to South America but has been widely spread as a contaminant of lucerne seed (Costea *et al.*, 2015). It is recorded in South Africa and Northern Africa (Algeria and Morocco), but to our knowledge it is not reported anywhere as a current weed problem in Africa.

4.7 Comparison of *Cuscuta* spp. and *Cassytha* spp.

One of the most striking examples of parallel evolution or homoplasy in angiosperms is the pair of obligate hemiparasites, *Cuscuta* spp. and *Cassytha* spp. Both are essentially leafless vines, similar in colour and repeatedly confused in the literature. For a discussion of the difficulties and distinctions, see Chapter 3, this volume.

4.8 Control

Control of dodder poses several unique challenges, especially regarding the biology of the seeds. The serious parasitic weeds in the genera *Striga* and *Orobanche* all require host stimulants to germinate. Dodder seeds have no such

Fig. 4.11. (A) *Cuscuta kilimanjari* on bougainvillaea, Arusha, Tanzania. (B) *C. kilimanjari* on *Manihot esculenta*, Rwanda.

requirement and are easy to germinate and in fact are used in teaching plant pathology (Nickrent and Musselman, 2016), although the seedcoats are tough and may require scarification to germinate. In nature, only a small percentage of the usually very large number of seeds in a field germinate. For example, one plant of *C. campestris* can produce 16,000 seeds (Costea and Tardif, 2005), only a fraction of which will germinate during the season they are shed.

In the most comprehensive review of control in dodder, Dawson *et al.* (1994) noted ten control measures: prevention, physical removal of the parasite, biological control, resistant host varieties, cultural control, mechanical control, fire and heat, foliage applied herbicides (selective and non-selective), and indirect chemical control. Lanini and Kogan (2005) provided a helpful overview and update of recent control methods. Numerous web sites include the same or similar approaches.

4.8.1 Prevention

Prevention includes the use of clean seed. Because dodder produces its seeds in both physical and temporal proximity to the host crop seed, there is a greater likelihood of parasite seed being harvested with the crop. Care should be taken to avoid seed spread by mud on the hoofs of animals as well as on equipment.

4.8.2 Physical removal

For the African smallholder farmer, the physical removal of dodders may be the only option for control. This requires careful monitoring of the crop so that young dodder plants are eliminated as soon as they appear. Dodder that has been removed should be burned as soon as possible. The parasite is fleshy, and fruits can develop and seeds mature from food stored in the stems.

4.8.3 Biological and genetic control

Biological control for dodders is very limited and has not contributed significantly to meaningful control. The most widely used agent is the fungus *Colletotrichum gloeosporioides*. Likewise, development of resistant host varieties has been meagre (reviewed in Kaiser *et al.*, 2015).

4.8.4 Cultural control

Exploiting the host selection adaptation of *Cuscuta* spp. is a cost-effective cultural means of control. For example, *C. campestris* does not attack monocots (except for onions, *Allium cepa*, and asparagus, *Asparagus officinalis*) so a crop rotation with cereals in an infested field would eliminate a host. Another example is a field of flax (*Linum usitatissimum*) infected with *C. epilinum*. Planting with a different crop than flax would deprive the parasite of a host. If these sorts of crop rotations are used, care is needed to ensure that dodder does not become established on weeds in the field.

Because there is a short phase between dodder germination and it being fully established on its hosts, control measures have centred on this brief autotropism in the parasite's life cycle. Heavy shade suppresses dodder seedlings, so planting a crop that would produce shade will reduce the number of dodder seedlings attaching to the host. Dodders need water for seed imbibition so scheduling irrigation can be a means of control. In mechanized fields, shallow ploughing will bury the unattached seedlings. This requires careful monitoring of germination. Burning will destroy most of the dodder seed in the stubble of a field.

4.8.5 Chemical control

There is a considerable corpus of literature on chemical control of dodder in mechanized agriculture (Parker and Riches, 1993; Dawson *et al.*, 1994; Costea and Tardif, 2005). Consequently, chemical control methods have been well developed for *Cuscuta campestris* (see Goldwasser *et al.*, 2012, and numerous websites for updates). Two broad approaches are selective and non-selective herbicides. Selective herbicides can be soil-applied to kill the dodder seedlings in the soil or soon after they emerge. Chlorpropham and imazaquin are two herbicides that have shown efficacy. Foliage-applied herbicides are translocated to kill the parasite, although some haustoria may remain in the host and grow. The most effective control in lucerne is glyphosate at 75 g ha^{-1}. Non-selective foliage-applied herbicides such as paraquat effectively kill the parasite but also severely damage the host. Most of the research on dodder control has been carried out in temperate regions with only a few investigations in tropical agriculture. However, control of *C. campestris* under tropical conditions in (ornamental) woody plants was studied by Hock *et al.* (2008). They found good control with glyphosate at 140 g ha^{-1}.

Lastly, indirect chemical control is achieved when annual weeds are controlled, thus reducing the secondary hosts, sometimes referred to as 'nurse plants'; these are not preferred hosts but rather plants that can provide temporary host services until the preferred host is attacked. It is important to distinguish between host range (the plants that the parasite favours) and host preference (the plant that provides optimal support for the dodder).

References

Beliz T.C. (1986) A revision of *Cuscuta* section *Cleistogrammica* using phentic and cladistic analyses with a comparison of reproductive mechanisms and host preferences in species from California, Mexico, and Central America. PhD dissertation, University of Berkeley, California.

Bennett, C.W. (1940) Acquisition and transmission of viruses by dodder (*Cuscuta subinclusa*). *Phytopathology* 30, 649–656.

Bennett, C.W. (1944) Studies of dodder transmission of plant viruses. *Phytopathology* 34, 905–932.

Birschwilks M., Haupt S., Hofius D. and Neumann S. (2006) Transfer of phloem-mobile substances from the host plants to the holoparasite *Cuscuta* sp. *Journal of Experimental Botany* 57, 911–921.

Chak, W.H., Tennakoon, K.U. and Musselman, L.J. (2010) The first report about dodders, the angiosperm parasitic genus *Cuscuta* (Yuncker) in Brunei Darussalam: a mystifying occurrence. *Folia Malaysiana* 11, 13–24.

Costea, M. and Tardif, F.J. (2005) The biology of Canadian weeds. 133. *Cuscuta campestris* Yuncker, *C. gronovii* Willd. ex Schult., *C. umbrosa* Beyr. ex Hook., *C. epithymum* (L.) L. and *C. epilinum* Weihe. *Canadian Journal of Plant Science* 86, 293–316.

Costea, M., Garcia, M.A. and Stefanović, S. (2015) A phylogenetically based infrageneric classification of the parasitic plant genus *Cuscuta* (dodders, Convolvulaceae). *Systematic Botany* 40, 269–285.

Dawson, J.H., Musselman, L.J., Dörr, I. and Wolkswinkel, P. (1994) Biology and control of *Cuscuta. Reviews of Weed Science* 6, 265–317.

Dobignard, A. and Chatelain, C. (2011) *Convolvulaceae.* In: Dobignard, A. and Chatelain, C. (eds) *Index synonymique de la Flore d'Afrique du nord*. Vol. 3. Conservatoire et Jardin Botanique de la Ville de Genève, Switzerland, pp. 331–351.

Donnapee, S. and Li, J. (2014) *Cuscuta chinensis* Lam.: a systematic review on ethnopharmacology, phytochemistry and pharmacology of an important traditional herbal medicine. *Journal of Ethnopharmacology* 157, 292–308.

Ghantous, K.M. and Sandler, H.A. (2012) Mechanical scarification of dodder seeds with a handheld rotary tool. *Weed Technology* 26, 485–489.

Goldwasser, Y., Miranda Sazo, M.R. and Lanini, W.T. (2012) Control of field dodder (*Cuscuta campestris*) parasitizing tomato with ALS-inhibiting herbicides. *Weed Technology* 26, 740–746.

Heide-Jørgensen, H. (2008) *Parasitic Flowering Plants*. Brill, Leiden, The Netherlands.

Hock, S.M., Wiecko, G. and Knezevic, S.Z. (2008) Glyphosate dose affected control of field dodder (*Cuscuta campestris*) in the tropics. *Weed Technology* 22, 151–155.

Kaiser, B., Vogg, G., Furst, U.D. and Albert, M. (2015) Parasitic plants of the genus *Cuscuta* and their interaction with susceptible and resistant host plants. *Frontiers in Plant Science* 6: 45.

Kelly, C.K. (1992) Resource choice in *Cuscuta europaea. Proceedings of the National Academy of Sciences USA* 89, 12194–12197.

Knepper, D.A., Creager, R.A. and Musselman, L.J. (1990) Identifying dodder seed as contaminants in seed shipments. *Seed Science and Technology* 18, 731–741.

Kuijt, J. (1969) *Parasitic Flowering Plants*. University of California Press, Berkeley, California.

Lanini, W.T. and Kogan, M. (2005) Biology and management of *Cuscuta* in crops. *International Journal of Agriculture and Natural Resources* 32, 127–141.

Masanga, J., Mwangi, B.N., Kibet, W., Sagero, P., Wamalwa, M. *et al.* (2021) Physiological and ecological warnings that dodders pose an exigent threat to farmlands in Eastern Africa. *Plant Physiology* 185, 1457–1467.

Musselman, L.J. (1986) Parasitic weeds and their impact in South-West Asia. *Proceedings of the Royal Society of Edinburgh, Section B: Biologival Sciences* 89, 283–288.

Musselman, L.J. (2019) *Parasitic Weeds in African Agriculture*. ECHO Technical Note 94, 12 pp.

Musselman, L.J., Aggour, M. and Abu-Sbaieh, H. (1989) Survey of parasitic weed problems in the West Bank and Gaza Strip. *Tropical Pest Management* 35, 30–33.

Muthaiyan, M.C. (2009) *Principles and Practices of Plant Quarantine*. Allied Publishers, New Delhi.

Nickrent, D.L. and Musselman, L.J. (2016) Parasitic plants. In: Ownley, B.H. and Trigano, R.N. (eds) *Plant Pathology Concepts and Laboratory Exercises, 3rd edn*. CRC Press, Boca Raton, Florida, pp. 277–288.

Olszewski, M., Dilliott, M., Garcia-Ruiz, I., Bandarvandi, B. and Costea, M. (2020) *Cuscuta* seeds: diversity and evolution, value for systematics/identification and exploration of allometric relationships. *PloS One* 15: e0234627.

Parker, C. and Riches, C.R. (1993) *Parasitic Plants of the World: Biology and Control*. CAB International, Wallingford, UK.

Ren, Z., Zagortchev, K., Ma, J., Yan, M. and Li, J. (2020) Predicting the potential distribution of the parasitic *Cuscuta chinensis* under global warming. *BMC Ecology* 20: 28.

Stolarz, M. (2009) Circumnutation as a visible plant action and reaction physiological, cellular and molecular basis for circumnutations. *Plant Signaling and Behavior* 4, 380–387.

Yu, H., Liu, J., He, W., Miao, S. and Dong, M. (2011) *Cuscuta australis* restrains three exotic invasive plants and benefits native species. *Biological Invasions* 13, 747–756.

Yuncker, T.G. (1932) The genus *Cuscuta*. *Memoirs of the Torrey Botanical Club* 18, 113–331.

Part II Root Parasites

As the name suggests, root parasites attack the roots of host plants. Root parasites are the most diverse parasites in terms of germination requirements, damage to subsistence crops and degree of photosynthetic dependence on the host. Root parasites comprise the most serious weeds, with severe economic and food-security impacts. They may contain chlorophyll (hemiparasites) or lack chlorophyll (holoparasites) and can also be distinguished by dependence on host root stimulation for seed germination (obligate parasites) and independence of these factors (facultative parasites).

Facultative parasitic plants do not depend on the presence of a host plant for seed germination and further growth and development. When after germination and seedling development a suitable host plant is found, they can turn into parasites. Facultative parasites benefit considerably from parasitism, as evident from increased biomass and seed production compared with independently growing plants. The most prominent weedy member of facultative parasites is rice vampire weed, which is widely distributed in Africa and primarily found on waterlogged soils.

Obligate hemiparasitic plants are parasites that contain chlorophyll but require a host to survive. Their malign behaviour is masked by their green appearance. This belies the seriousness of their damage. Indeed, obligate hemiparasites comprise some of the most serious parasitic weeds in Africa, causing millions of dollars of damage each, and under some conditions are a major cause of famine because several attack subsistence crops. Here we emphasize witchweeds, a group of parasites causing damage throughout the African continent, especially in semi-arid regions. One species infests legume crops, the others only grasses including cereal crops. Found throughout most of Africa, witchweed has its greatest effect in the semi-arid tropics. Relatives include rice vampire weed and alectra, which causes serious damage to legume crops. We also discuss some witchweeds and relatives that are not presently serious problems, but which should be recognized as having potential to become a problem.

Holoparasitic plants are strikingly different from the surrounding vegetation because they totally lack chlorophyll and therefore are always closely associated with a green host plant. Because they are totally dependent upon their host, they lack typical leaves and emerge only to flower. By that time, the host has been damaged before the farmer could intervene. Like witchweed, these parasites will not germinate without host influence. Most significant of these is broomrape which, unlike witchweed, does not parasitize cereals but can be devastating in vegetable crops. Although witchweed is largely tropical in distribution, broomrape favours cooler temperatures, and hence is usually only a problem at higher altitudes, or greater latitudes in Africa.

The diversity of root parasites includes permutations of each of these features. Important weedy representatives are presented in the following chapters, starting with the facultative parasites (rice vampire weed, Chapter 5, and buchnera, Chapter 6), then the obligate hemiparasites (witchweed, Chapter 7, and alectra, Chapter 8) and the holoparasites (broomrape, Chapter 9, and thonningia, Chapter 10). Lastly, we cover a few relatively unknown species of facultative or hemiparasitic nature (Chapter 11).

5 Rice Vampire Weed

Abstract

Rhamphicarpa fistulosa, known by the common name rice vampire weed, is the most problematic and widespread species among the facultative parasitic weeds in Africa. It is reported in 37 countries. Until recently, it was considered a parasitic weed of minor importance, possibly because it was notoriously overlooked. Another reason for a change in status from minor to major parasitic weed is an increase in rice production in rainfed lowlands where *R. fistulosa* thrives. In these environments, not many food crops other than rice can be grown. Rice is therefore the most important host, the source of its common name. The host range includes other cereal crops such as sorghum and maize, but dicots such as soybean and groundnut have also been reported as hosts. The weed can germinate and grow independently, but significantly increases its reproductive output when parasitizing a suitable host. Host growth and seed production are therefore greatly reduced and can even be arrested by *R. fistulosa* parasitism, leading to average crop yield losses of 24%–73% and estimated annual economic losses of around US$82 million. Farmers often have little knowledge and means to control this parasitic weed other than by applying post-emergence herbicides or hand weeding.

5.1 Introduction

A parasitic weed that has attracted increased attention over the past decade is rice vampire weed, *Rhamphicarpa fistulosa* (Hochst.) Benth. Little known by local agricultural extension workers and researchers, this species has become a widespread problem of rainfed cereal crops, primarily lowland rice with limited control of water, in sub-Saharan Africa (Rodenburg *et al.*, 2015a). It has probably been overlooked and underestimated for a long time because of the inconspicuous appearance of the species, with opposite leaves divided into needle-like segments and slender white flowers with a long narrow tube topped with five lobes that remain closed during daytime (Fig. 5.1). Another reason is likely the suboptimal and sometimes dysfunctional communication between

 Parasitic Plants in African Agriculture (L.J. Musselman and J. Rodenburg)
DOI: 10.1079/9781789247657.0005

Fig. 5.1. The unobtrusive appearance of the facultative parasitic plant *Rhamphicarpa fistulosa*. (A) The whole plant, with its rice host in the background. (B) Branch of a vegetative plant showing the needle-like leaf segments. (C) Close-up of a flower during daytime. (D) Close-up of a flower during the night. (E) Reproductive part of the plant, with developing capsules.

smallholder farmers and extension services. In Benin for instance, farmers struggled with the weed for many years, and although it was observed by local weed scientists (Gbèhounou and Assigbé, 2003), nearby extension agents were largely unaware of the problem (Rodenburg *et al.*, 2015b). Until recently, there was also a paucity of understanding of the importance of this weed and its biology, ecology and effective management strategies (Tippe *et al.*, 2017a) throughout the whole range of stakeholders, i.e. farmers, extension agents, researchers and crop protection services (Schut *et al.*, 2015a, b).

Across Africa, this parasitic weed occurs in lowland areas where farmers have no control over water (Gworgwor *et al.*, 2001; Rodenburg *et al.*, 2011; Houngbedji *et al.*, 2014; N'Cho *et al.*, 2014; Rodenburg *et al.*, 2016a). Surveys conducted in Benin, Côte d'Ivoire and Nigeria showed that 13%–48% of these lowlands were infested (Gworgwor *et al.*, 2001; N'Cho *et al.*, 2014; Rodenburg *et al.*, 2011). Based on these data and a spatial modelling exercise, the mean incidence of *R. fistulosa* in rainfed lowland areas used for rice production across sub-Saharan Africa is estimated at 6% (Rodenburg *et al.*, 2016b). Obviously, the yield losses caused by this weed vary, depending on local biophysical conditions, such as soil fertility and water management, and agronomic factors, such as weeding frequency, crop species and cultivar choice. Reported rice yield losses caused by *R. fistulosa* are estimated to be 24%–73% (Rodenburg *et al.*, 2011; N'Cho *et al.*, 2014; Rodenburg *et al.*, 2016a). On a regional scale, annual (milled) rice production losses are estimated at 0.2 million tonnes, worth around US$82 million (Rodenburg *et al.*, 2016b).

Rhamphicarpa fistulosa is different from most of the other parasites described in this book as it is a facultative parasitic plant, meaning that it can germinate and complete its life cycle entirely in the absence of a host (Ouédraogo *et al.*, 1999). The other African facultative parasitic weeds are *Buchnera hispida* (see Chapter 6, this volume) and, presumably, *Sopubia parviflora*, *Micrargeria filiformis* and *Thesium humile* (see Chapter 11, this volume). The host-independence characteristic may have little relevance under natural conditions, as in natural vegetation facultative parasitic plants are usually assumed to be connected to hosts (Estabrook and Yoder, 1998), but for the management of this species this has important implications. When the parasite cannot connect to a host, it is stunted and seed production is much diminished compared with that of parasitizing counterparts (Kabiri *et al.*, 2016). Hence, avoiding host contact, for example by spatial or temporal separation (Tippe *et al.*, 2017b) or by making use of resistant crop varieties (Rodenburg *et al.*, 2016a), may reduce reproduction rates of the weeds and mitigate weed problems in subsequent years.

5.2 Taxonomy

The taxonomy of *R. fistulosa* is described by Hochstetter (1841) and Bentham (1846) and revised by Hansen (1975) and Mielcarek (1996). A synopsis of its nomenclature is published in Rodenburg *et al.* (2015a) showing several synonyms, for example *Macrosiphon elongatus* (Hochst.) and *Rhamphicarpa australiensis* Steenis, and vernacular names, for example Kayongo, Mbosyo, Mulungi, Angamay (Eastern Africa), Otcha, Corico, Efri and Loha Soukoh (Western Africa). Detailed botanical and ecological descriptions of the species are given in Hansen (1975) and Philcox (1990). Plants are usually short (<120 cm), slender and erect, with glabrous stems and leaves. Flowers are white with long tubular corollas and asymmetrical pea-sized seed capsules with beaks. Seeds are tiny and light (0.2 × 0.55 mm; 0.011 mg) and oval shaped with a corrugated surface (Mielcarek, 1996; Ouédraogo *et al.*, 1999; Rodenburg *et al.*, 2011). The parasite has a reduced root system (Rodenburg and Bastiaans, 2019). The typical habitat is open grasslands and seasonally inundated wetlands (Gledhill, 1970; Hansen, 1975). It is mostly observed in rainfed lowland rice fields, but also in better-drained upland areas where maize or sorghum is grown (Ouédraogo *et al.*, 1999; Rodenburg *et al.*, 2015a).

5.3 Distribution

The species is reported in 37 countries in Africa south of the Sahara, from Senegal to Madagascar and from Sudan to South Africa (Fig. 5.2). The most

Fig. 5.2. *Rhamphicarpa fistulosa* distribution within Africa (mapped distribution depends on extent of collections). Data from: Kew Royal Botanic Gardens; South African National Biodiversity Institute; Missouri Botanical Garden (MBG); MBG's TROPICOS database; Muséum National d'Histoire Naturelle (SONNERAT); l'Herbier de Parc Botanique et Zoologique de Tsimbazaza Antananarivo, Madagascar; l'Herbier du Département de Biologie et Ecologie Végétales (DBEV), Université d'Antananarivo; National Herbarium of Rwanda; Herbarium of the Department of Botany, University of Dar es Salaam; East African Herbarium, National Museums of Kenya. Black dots indicate georeferenced records and non-grey colours indicate all countries where the species has been observed.

affected countries are Burkina Faso, Cameroon, Côte d'Ivoire, Guinea, Mali, Nigeria and Sierra Leone in Western Africa, and Madagascar, Tanzania and Uganda in Eastern Africa. Based on 419 available observation points distributed over these 37 countries, and a regional map of rainfed rice cultivation areas, the total area of infested rice fields was conservatively estimated to be around 455,000 ha (Rodenburg *et al.*, 2016b). Estimates for other food crops, such as sorghum and maize are not available.

5.4 Biology

As noted, *Rhamphicarpa fistulosa* is a facultative root parasite. The species can germinate, grow and complete its life cycle like any normal, autotrophic plant. Seeds require light and moist conditions to germinate. Kabiri *et al.* (2016) showed that the germination of *R. fistulosa* was not influenced by natural (host-derived) or synthetic (e.g. GR24) germination stimulants required for germination of obligate root parasites of the same family (e.g. *Striga* spp.; see Chapters 7 and 13, this volume). After germination, a seedling develops with chlorophyll enabling the plant to assimilate CO_2 and grow independently. A plant of *R. fistulosa* only becomes parasitic once its roots form haustoria. This means that parasitism often does not occur in the first 4–6 weeks of the life cycle (Kabiri *et al.*, 2017). Once the roots of the parasite encounter roots of a suitable host, lateral haustoria develop that connect to and penetrate the host root, forming xylem–xylem connections (Neumann *et al.*, 1998, 1999) allowing the parasite to extract water and nutrients and possibly metabolites (through phloem bridges) from its host.

Upon successful infection of a suitable host, *R. fistulosa* plants show steeply increased biomass accumulation, both above- and belowground (Rodenburg and Bastiaans, 2019), as well as an increase in capsule numbers with enhanced seed production (Ouédraogo *et al.*, 1999). By parasitizing a suitable host, this facultative parasite can increase its seed production four- to 13-fold (Kabiri *et al.*, 2016; Rodenburg and Bastiaans, 2019); hence parasitism has an obvious advantage for the species. The host, on the other hand, can be severely affected by parasitism of *R. fistulosa* (Fig. 5.3).

The parasite can manipulate its host so that the host plant ends up producing assimilates solely for its unwanted guest with growth of the host plant arrested. Kabiri *et al.* (2017) have shown that the parasite increases the root:shoot ratio of rice plants upon infection, mainly caused by very strong (22%–71%) aboveground biomass reductions. The observed aboveground growth reduction

Fig. 5.3. Rice (variety IR64) growing without (left) and with (centre) the parasite (*Rhamphicarpa fistulosa*), and the parasite growing with (centre) and without (right) the host plant (rice). Note the severe reduction in rice biomass and reproduction due to the presence of the parasite and the increase in parasite biomass and reproduction due to the presence of a host.

of the host is a result of a combination of the parasite-induced reductions in light interception and light-use efficiency (Kabiri *et al.*, 2017), which in turn is caused by steeply reduced leaf-level photosynthesis rates (Kabiri *et al.*, 2017).

Although *R. fistulosa* is reported to be allogamous and dependent on specific pollinators, such as night moths (Cissé *et al.*, 1996; Ouédraogo *et al.*, 1999), we observed seed production in experimental greenhouses in Europe in the absence of pollinators (J. Rodenburg, personal observation). *Rhamphicarpa fistulosa* is a prolific seed producer (~1000 seeds per plant), in particular when it is parasitic, but its seed viability in the soil is relatively short at around 1 year (Gbèhounou and Assigbé, 2004). Seeds remain viable for longer times (>5 years) if stored under dry and shaded conditions (J. Rodenburg, personal observation). Short seed longevity seems to be a characteristic trait of facultative parasitic plants (Kelly, 1989; Strykstra *et al.*, 1996; Bekker and Kwak, 2005) and one of the traits that render a facultative life-cycle strategy more successful than an obligate strategy (Bastiaans and Rodenburg, 2019).

5.5 Hosts

The host range of *R. fistulosa* is broad but noted mainly on grasses (Parker, 2013). Apart from members of the Poaceae, it has been reported on weed species of Cyperaceae, Leguminosae and Labiatae (Bouriquet, 1933) but this needs confirmation (Houngbedji *et al.*, 2016). Cultivated plant species – cereal crops such as rice, maize, millet and sorghum – are known hosts of *R. fistulosa* (Cissé *et al.*, 1996; Ouédraogo *et al.*, 1999), but a range of non-cereal crop species have also been reported, such as soybean (*Glycine max*), groundnut (*Arachis hypogaea*) and the leafy vegetable crop tossa jute (*Corchorus olitorius*), popular in Western Africa (Houngbedji and Gibot-Leclerc, 2015).

5.6 Control

The broad host range essentially excludes the use of crop rotations and intercropping as an effective control measure. This is an often-proposed strategy to reduce the seed bank increments in crop fields infested by *Striga* spp. (e.g. Van Mourik *et al.*, 2008). Another control strategy against *Striga* spp. is the application of organic or mineral fertilizers (e.g. Farina *et al.*, 1985; Raju *et al.*, 1990; Jamil *et al.*, 2012). For *R. fistulosa*, this does not reduce but rather stimulates parasite growth and biomass accumulation (Tippe *et al.*, 2020) and thereby increases seed production (Rodenburg and Bastiaans, 2019).

On the other hand, because of the delay before host contact by young *R. fistulosa*, plants can be removed by hand or hoe weeding, or killed by post-emergence herbicides, such as 2,4-dichlorophenoxyacetic acid (2,4-D) (Gbèhounou and Assigbé, 2003), before severe damage to the crop has occurred. An early (2 weeks after emergence of *R. fistulosa* plants) single application of 2,4-D at 820 g ha^{-1}, or a repeated application at 360 g ha^{-1} (with a 7 day interval), has proven very effective (Ouédraogo *et al.*, 2017).

The applicability of post-emergence herbicides against *R. fistulosa* is a clear difference from obligate parasitic weeds, such as *Striga* spp., which cause a critical and irremediable negative impact on their hosts when they are still belowground (e.g. Gurney *et al.*, 1995) and when intervention measures cannot be applied.

A relatively simple control measure used at no additional cost for the farmer is early sowing. In southern Tanzania, experiments in crop establishment timing were conducted in rice fields infested with *R. fistulosa* nearby those infested with *Striga asiatica*; these show that when the rice crop is sown earlier than usual, the *R. fistulosa* infestation is greatly reduced (Tippe *et al.*, 2017b). Here again the control practice for *R. fistulosa* contrasts with that of *S. asiatica*, which was managed by delayed sowing.

Another low-cost control measure is the use of resistant or tolerant cultivars. For rice, several such cultivars have been identified. Lowland NERICA (New Rice for Africa) cultivars (e.g. *NERICA-L-40* and *-31*) maintain reasonably high crop yields and low parasite-infection levels under *R. fistulosa*-infested field conditions (Rodenburg *et al.*, 2016a). Interestingly, even the rice cultivar *Supa India*, the popular choice of farmers in Tanzania, maintains good yields under infested conditions, but the infection levels of this cultivar are high. Tanzanian rice farmers often prefer this local cultivar over most of the modern ones, because it combines reasonable yields with a desirable grain quality and therefore has good marketability. They do, however, appreciate cultivars that are early maturing or resistant (Tippe *et al.*, 2017a).

The use of herbicides (e.g. 2,4-D) and hand weeding remain the two most popular control measures currently used by rice farmers in southern Tanzania and other infested areas in Africa.

5.7 Comparison of *R. fistulosa* with Other *Rhamphicarpa* Species and *Striga* Species

The other *Rhamphicarpa* species that overlap geographically with *R. fistulosa* are *R. elongata* (Hochst) O.J. Hansen, observed in Chad, Sudan and Ethiopia (GBIF, 2022); *R. veronicaefolia*, observed in Kenya and Tanzaia; *R. capillacea* A. Raynal, restricted to Central Africa (Democratic Republic of the Congo (DRC) and the Central African Republic) and sharing a similar (wetland) environmental preference (Raynal, 1970); and *R. brevipedicellata* O.J. Hansen, which seems restricted to Southern Africa (Zambia, Botswana, Namibia, Eswatini, South Africa), although also one non-georeferenced observation has been recorded from DRC, according to a specimen at Meise Botanic Garden Herbarium (GBIF, 2022) (Fig. 5.4). These are all presumed facultative parasites, although none of them has been reported as a weed problem.

Due to the similarity in affected cropping systems, farm and farmer typology, *R. fistulosa* is compared to species of the genus *Striga*, most notably the important weeds of tropical cereals such as *S. asiatica*, *S. aspera* and *S. hermonthica* (discussed in Chapter 7, this volume). Like *R. fistulosa*, the *Striga* species are root-parasitic plants that negatively impact hosts and consequently cause

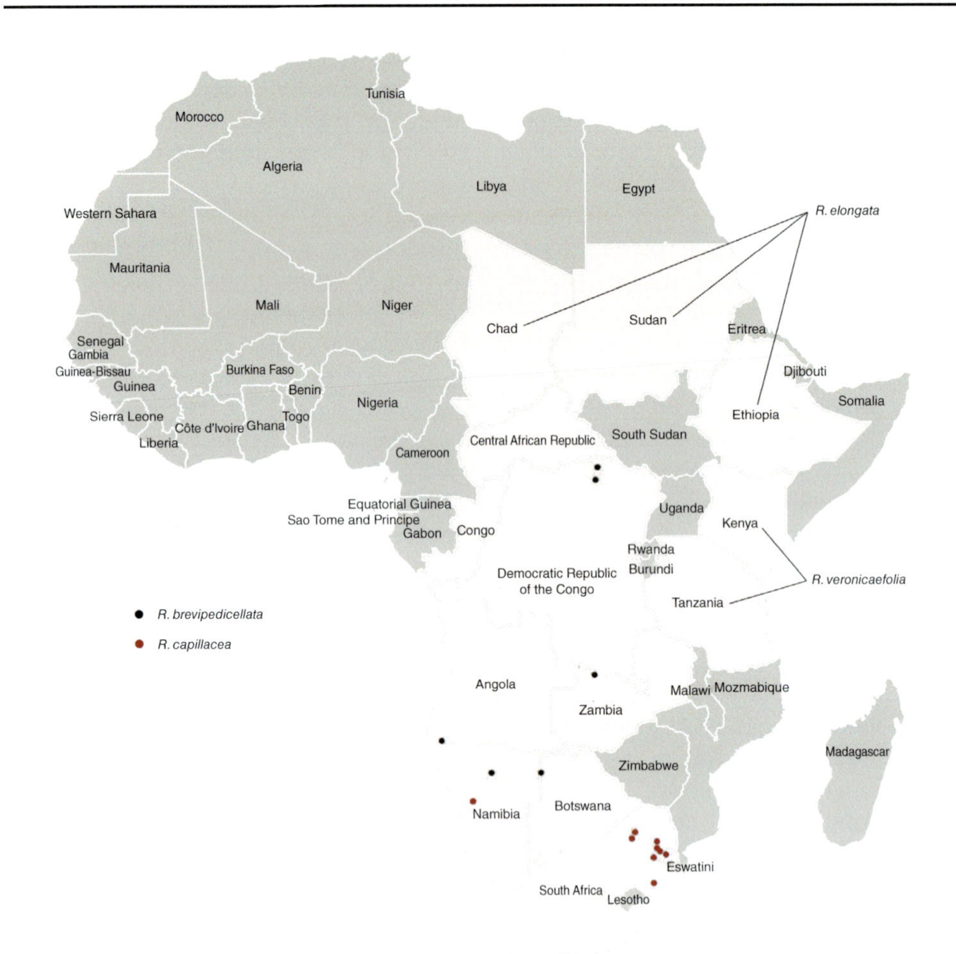

Fig. 5.4. *Rhamphicarpa brevipedicellata*, *R. capillacea*, *R. veronicaefolia* and *R. elongata* distribution within Africa (mapped distribution depends on extent of collections; no specific geographic references for *R. veronicaefolia* and *R. elongata*). Data from Global Biodiversity Information Facility: GBIF.org (accessed 28 August 2022), GBIF Occurrence Download https://doi.org/10.15468/dl.28g9wv, https://doi.org/10.15468/dl.3ckxf2 and https://doi.org/10.15468/dl.fecmqf. Black and red dots indicate georeferenced records from GBIF and non-grey colours indicate all countries where the species has been observed.

large cereal crop yield losses in Africa (e.g. Parker, 2012) and they primarily affect smallholder farmers (e.g. Debrah, 1994; Scheiterle *et al.*, 2019; Wang *et al.*, 2019). Farmers do clearly struggle to differentiate between different species of parasitic weeds damaging their (cereal) crops. This is evident from surveys conducted in Senegal, Côte d'Ivoire and Uganda when we asked rice farmers whether they suffered from *Striga* spp. in their crops and then followed this up by field visits. Several of the farmers who answered affirmatively to the question were struggling with *R. fistulosa* infestations rather than *Striga* spp. (Fig. 5.5; J. Rodenburg, personal observation). In Uganda, farmers referred to *R. fistulosa* as 'Striga of rice' and they used a similar local vernacular name for

Fig. 5.5. Parasite effects on the host. (A) Rice crop (green plants) infested by *Rhamphicarpa fistulosa* (reddish plants), viewed from above. (B) Farmer's field in Benin with a weeded area of rice in the centre and unweeded parts at both sides completely infested by *R. fistulosa*. (C) Close-up of seed capsule producing *R. fistulosa* plants in a rice crop in Benin. (D) Rice farmer in Casamance, Senegal, showing uprooted *R. fistulosa* plants with her infested field in the background. (E) Rice farmer and extension agent in Tsiroanomandidy (Bongolava Region), Madagascar, showing uprooted *R. fistulosa* plants from the field. (Photo E courtesy of Mamadou Cissoko.)

both species (i.e. *Kayongo*). Also in central Benin, the vernacular names for *R. fistulosa* and *Striga* spp. (i.e. *Otcha, Do, Corico, Efri*) were the same (Rodenburg *et al.*, 2015a).

The main difference between *R. fistulosa* and *Striga* spp. is the type of parasitism; they are all root hemiparasites, but *R. fistulosa* is facultative whereas all *Striga* spp. are obligate parasites, meaning they are dependent upon a host for underground stages of their life cycle (including seed germination). A second important difference concerns environmental adaptations determining their main habitats. Whereas *R. fistulosa* has a broad environmental adaptation, its preferred habitat is seasonally flooded wetlands and their hydromorphic fringes where lowland rice is the only staple food crop that can be grown. *Striga* spp. occur primarily on free-draining upland soils where sorghum, millet, maize and upland rice are cultivated (e.g. Johnson *et al.*, 1997; Gworgwor *et al.*, 2001; Kamara *et al.*, 2014). Surveys in rainfed rice areas in southern Tanzania, where both types of parasitic weeds occur in rice fields, have shown that this environmental distinction between *R. fistulosa* and *Striga asiatica* is rather strict; along the lowland–upland continuum, *R. fistulosa* was primarily found in fields at lower positions and the hydromorphic transition zones, and *S. asiatica* primarily on the upper crests, and in only one of the surveyed fields were both species observed (Kabiri *et al.*, 2015).

The differences in biology (mainly host dependence) and ecology between the two types of parasitic weeds have important implications for the design of effective control strategies, as discussed in the previous section.

References

Bastiaans, L. and Rodenburg, J. (2019) Facultative parasitism: an evolutionary precursor of complete parasitism or an effective strategy in its own right? *15th World Congress on Parasitic Plants*, 30 June–5 July 2019, Amsterdam, The Netherlands. Available at: www.wcpp2019.org/wp-content/uploads/2019/07/WCPP2019_Conference_book_final.pdf (accessed 13 May 2023).

Bekker, R.M and Kwak, M.M. (2005) Life history traits as predictors of plant rarity, with particular reference to hemiparasitic Orobanchaceae. *Folia Geobotanica* 40, 231–242.

Bentham, G. (1846) Ordo CXLII. Scrophulariaceae. In: De Candolle, A. (ed.) *Prodromus Systematis Naturalis Regni Vegetabilis.* Paris Sumptibus Sociorum Treuttel et Würtz 10, pp. 186–586.

Bouriquet, G. (1933) Une Scrophulariacée parasite du riz à Madagascar. *Revue de Pathologie Végétale et d'Entomologie Agriculture de France* 20, 149–151.

Cissé, J., Camara, M., Berner, D.K. and Musselman L.J. (1996) *Rhamphicarpa fistulosa* (Scrophulariaceae) damages rice in Guinea. In: Moreno, M.T., Cubero, J.I., Berner, D.K., Joel, D., Musselman, L.J. and Parker, C. (eds) *Advances in Parasitic Plant Research: 6th Parasitic Weeds Symposium.* Cordoba, Spain, pp. 518–520.

Debrah, S.K. (1994) Socio-economic constraints to the adoption of weed control techniques: the case of *Striga* control in the West African semi-arid tropics. *International Journal of Pest Management* 40, 153–158.

Estabrook, E.M. and Yoder, J.I. (1998) Plant–plant communications: rhizosphere signaling between parasitic angiosperms and their hosts. *Plant Physiology* 116, 1–7.

Farina, M.P.W., Thomas, P.E.L. and Channon, P. (1985) Nitrogen, phosphorus and potassium effects on the incidence of *Striga asiatica* (L.) Kuntze in maize. *Weed Research* 25, 443–447.

Gbèhounou, G. and Assigbé, P. (2003) *Rhamphicarpa fistulosa* (Hochst.) Benth. (Scrophulariaceae): new pest on lowland rice in Benin. Results of a survey and immediate control possibilities. *Annales des Sciences Agronomique du Bénin* 4, 89–103.

Gbèhounou, G. and Assigbé, P. (2004) A study on germination of seeds of *Rhamphicarpa fistulosa* (Hochst.) Benth., a new pest of rice. In: *4th International Weed Science Conference, 20–24 June 2004, Durban, South Africa*. International Weed Science Society, Haryana, India. Available at: www.iwss.info/proceedings-of-meetings.html (accessed 13 May 2023).

GBIF (2022) Global Biodiversity Information Facility. Available at: www.gbif.org (accessed 18 May 2023).

Gledhill, D. (1970) Vegetation of superficial ironstone hardpans in Sierra Leone. *Journal of Ecology* 58, 265–274.

Gurney, A.L., Press, M.C. and Ransom, J.K. (1995) The parasitic angiosperm *Striga hermonthica* can reduce photosynthesis of its sorghum and maize hosts in the field. *Journal of Experimental Botany* 46, 1817–1823.

Gworgwor, N.A., Ndahi, W.B. and Weber, H.C. (2001) Parasitic weeds of north-eastern region of Nigeria: a new potential threat to crop production. In: *The BCP Conference, 13–15 November 2001, Brighton, UK*. British Crop Protection Council, Farnham, UK, pp. 181–186.

Hansen, O.J. (1975) The genus *Rhamphicarpa* Benth. emend. Engl. (Scrophulariaceae). A taxonomic revision. *Botanisk Tidsskrift* 70, 103–125.
Hochstetter, C.H. (1841) Plantarum nubicarum nova genera. *Flora* 24, 369–384.
Houngbedji, T. and Gibot-Leclerc, S. (2015) First report of *Rhamphicarpa fistulosa* on peanut (*Arachis hypogaea*), soybean (*Glycine max*), and tossa jute (*Corchorus olitorius*) in Togo. *Plant Disease* 99, 1654–1655.
Houngbedji, T., Pocanam, Y., Shykoff, J., Nicolardot, B. and Gibot-Leclerc, S. (2014) A new major parasitic plant in rice in Togo: *Rhamphicarpa fistulosa*. *Cahiers Agricultures* 23, 357–365.
Houngbedji, T., Dessaint, F., Nicolardot, B., Shykoff, J.A. and Gibot-Leclerc, S. (2016) Weed communities of rain-fed lowland rice vary with infestation by *Rhamphicarpa fistulosa*. *Acta Oecologica* 77, 85–90.
Jamil, M., Kanampiu, F.K., Karaya, H., Charnikhova, T. and Bouwmeester, H.J. (2012) *Striga hermonthica* parasitism in maize in response to N and P fertilisers. *Field Crops Research* 134, 1–10.
Johnson, D.E., Riches, C.R., Diallo, R. and Jones, M.J. (1997) *Striga* on rice in West Africa; crop host range and the potential of host resistance. *Crop Protection* 16, 153–157.
Kabiri, S. Rodenburg, J., van Ast, A., Pflug, S., Kool, H. *et al.* (2021) Impact of the facultative parasitic weed *Rhamphicarpa fistulosa* (Hochst.) Benth. on photosynthesis of its host *Oryza sativa* L. *Journal of Plant Physiology* 262, 153438.
Kabiri, S., Rodenburg, J., Kayeke, J., Makokha, S., Msangi, S.H. *et al.* (2015) Can the parasitic weeds *Striga asiatica* and *Rhamphicarpa fistulosa* co-occur in rain-fed rice? *Weed Research* 55, 145–154.
Kabiri, S., Van Ast, A., Rodenburg, J. and Bastiaans, L. (2016) Host influence on germination and reproduction of the facultative hemi-parasitic weed *Rhamphicarpa fistulosa*. *Annals of Applied Biology* 169, 144–154.
Kabiri, S., Rodenburg, J., Van Ast, A. and Bastiaans, L. (2017) Slavery in plants: how the facultative hemi-parasitic plant *Rhamphicarpa fistulosa* can completely dominate its host. *Annals of Applied Biology* 171, 353–363.
Kamara, A.Y., Ekeleme, F., Jibrin, J.M., Tarawali, G. and Tofa, I. (2014) Assessment of level, extent and factors influencing *Striga* infestation of cereals and cowpea in a Sudan Savanna ecology of northern Nigeria. *Agriculture, Ecosystems & Environment* 188, 111–121.
Kelly, D. (1989) Demography of short-lived plants in chalk grassland. 1. Life-cycle variation in annuals and strict biennials. *Journal of Ecology* 77, 747–769.
Mielcarek, R. (1996) The Scrophulariaceae in the flora of Central Africa (excl. Lindernieae). *Fragmenta Floristica et Geobotanica* 41, 3–248.
N'cho, S.A., Mourits, M., Rodenburg, J., Demont, M. and Lansink, A.O. (2014) Determinants of parasitic weed infestation in rainfed lowland rice in Benin. *Agricultural Systems* 130, 105–115.
Neumann, U., Sallé, G. and Weber, H.C. (1998) Development and structure of the haustorium of the parasite *Rhamphicarpa fistulosa* (Scrophulariaceae). *Botanica Acta* 111, 354–365.
Neumann, U., Vian, B., Weber, H.C. and Sallé, G. (1999) Interface between haustoria of parasitic members of the Scrophulariaceae and their hosts: a histochemical and immunocytochemical approach. *Protoplasma* 207, 84–97.
Ouédraogo, O., Neumann, U., Raynal Roques, A., Sallé, G., Tuquet, C. *et al.* (1999) New insights concerning the ecology and the biology of *Rhamphicarpa fistulosa* (Scrophulariaceae). *Weed Research* 39, 159–169.
Ouédraogo, O., Kaboré, I., Kaboré, T. and Boussim, I.J. (2017) Effets de traitements herbicides sur *Rhamphicarpa fistulosa* (Hochst.) Benth. Une plante parasite facultative. *Journal of Applied Biosciences* 119, 11983–11992.
Parker, C. (2012) Parasitic weeds: a world challenge. *Weed Science* 60, 269–276.

Parker, C. (2013) The parasitic weeds of the Orobanchaceae. In: Joel, D.M., Gressel, J. and Musselman, L.J. (eds) *Parasitic Orobanchaceae: Parasitic Mechanisms and Control Strategies.* Springer, Berlin, pp. 313–344.

Philcox, D. (1990) Scrophulariaceae. *Flora Zambesiaca* 8, 1–179.

Raju, P.S., Osman, M.A., Soman, P. and Peacock, J.M. (1990) Effects of nitrogen, phosphorus and potassium on *Striga asiatica* (L.) Kuntze seed germination and infestation of sorghum. *Weed Research* 30, 139–144.

Raynal, A. (1970) Un nouveau *Rhamphicarpa* (Scrophulariaceae) d'Afrique Central. *Adansonia* 10, 329–332.

Rodenburg, J. and Bastiaans, L. (2019) How do fertilizers affect the facultative parasitic weed *Rhamphicarpa fistulosa*? In: *15th World Congress on Parasitic Plants*, 30 June–5 July 2019, Amsterdam, The Netherlands.

Rodenburg, J., Zossou-Kouderin, N., Gbèhounou, G., Ahanchede, A., Touré, A. *et al.* (2011) *Rhamphicarpa fistulosa*, a parasitic weed threatening rain-fed lowland rice production in sub-Saharan Africa – a case study from Benin. *Crop Protection* 30, 1306–1314.

Rodenburg, J., Morawetz, J.J. and Bastiaans, L. (2015a) *Rhamphicarpa fistulosa*, a widespread facultative hemi-parasitic weed, threatening rice production in Africa. *Weed Research* 55, 118–131.

Rodenburg, J., Schut, M., Demont., M., Klerkx, L., Gbèhounou, G. *et al.* (2015b) Systems approaches to innovation in pest management: reflections and lessons learned from an integrated research program on parasitic weeds in rice. *International Journal of Pest Management* 61, 329–339.

Rodenburg, J., Demont, M., Zwart, S.J. and Bastiaans, L. (2016a) Parasitic weed incidence and related economic losses in rice in Africa. *Agriculture, Ecosystems & Environment* 235, 306–317.

Rodenburg, J., Cissoko, M., Dieng, I., Kayeke, J. and Bastiaans, L. (2016b) Rice yields under *Rhamphicarpa fistulosa*-infested field conditions, and variety selection criteria for resistance and tolerance. *Field Crops Research* 194, 21–30.

Scheiterle, L., Haring, V., Birner, R. and Bosch, C. (2019) Soil, Striga, or subsidies? Determinants of maize productivity in northern Ghana. *Agricultural Economics* 50, 479–494.

Schut, M., Rodenburg, J., Klerkx, L., Hinnou, L.C., Kayeke J. *et al.* (2015a) Participatory appraisal of institutional and political constraints and opportunities for innovation to address parasitic weeds in rice. *Crop Protection* 74, 158–170.

Schut, M., Rodenburg, J., Klerkx, L., Kayeke, J., Van Ast, A. *et al.* (2015b) RAAIS: Rapid Appraisal of Agricultural Innovation Systems (Part II). Integrated analysis of parasitic weed problems in rice in Tanzania. *Agricultural Systems* 132, 12–24.

Strykstra, R.J., Bekker, R.M. and Verweij, G.L. (1996) Establishment of *Rhinanthus angustifolius* in a successional hayfield after seed dispersal by mowing machinery. *Acta Botanica Neerlandica* 45, 557–562.

Tippe, D.E., Rodenburg, J., Schut, M., Van Ast, A., Kayeke, J. *et al.* (2017a) Farmers' knowledge, use and preferences of parasitic weed management strategies in rain-fed rice production systems. *Crop Protection* 99, 93–107.

Tippe D.E., Rodenburg, J., Van Ast, A., Anten, N.P.R., Dieng, I. *et al.* (2017b) Delayed or early sowing: timing as parasitic weed control strategy in rice is species and ecosystem dependent. *Field Crops Research* 214, 14–24.

Tippe, D.E., Bastiaans, L., Van Ast, A., Dieng, I., Cissoko, M. *et al.* (2020) Fertilisers differentially affect facultative and obligate parasitic weeds of rice and only occasionally improve yields in infested fields. *Field Crops Research* 254: 107845.

Van Mourik, T.A., Bianchi, F.J.J.A., Van Der Werf, W. and Stomph, T.J. (2008) Long-term management of *Striga hermonthica*: strategy evaluation with a spatio-temporal population model. *Weed Research* 48, 329–339.

Wang, H., Snapp, S.S., Fisher, M. and Viens, F. (2019) A Bayesian analysis of longitudinal farm surveys in Central Malawi reveals yield determinants and site-specific management strategies. *Plos One* 14: e0219296.

6 Buchnera

Abstract
Another facultative root parasite genus is *Buchnera*. The only known weedy species in this genus is *B. hispida*. It is widely distributed across Africa, as evident from the records from 30 African countries, but only locally known as a weed. *Buchnera hispida* resembles *Striga* spp. in terms of its host range, environmental preference and appearance. Despite this resemblance, *B. hispida* is not yet known as an important weed in smallholder cropping systems in Africa. As a facultative parasitic weed, it does not require host root exudates for its germination.

6.1 Introduction

Buchnera spp. are a widespread group of native parasites with a weedy tendency but causing no reported significant damage. These native species prefer open, sunny, disturbed sites, expected behaviour for a plant that becomes an agrestal weed, that is, a plant adapted to cultivated land. About a hundred species of the genus have been named from Africa. Of these, the only species reported as an agricultural weed is *Buchnera hispida*. This is a widely distributed facultative hemiparasite. Nwoke and Okonkwo (1974) studied the parasitism of *B. hispida* and issued a warning that it posed a potential problem to crops. This was repeated in the 1990s by Raynal-Roques (1994), who also suggested it could become an important weed, but to date there are no signs this has happened. The few reports on *B. hispida* as a weed in cereal production systems are not recent and are limited to Western Africa, that is, Senegal, Mali and Nigeria (Nwoke and Okonkwo, 1974; Parkinson, 1989; Sallé *et al.*, 1994; Gworgwor *et al.*, 2001). Whether it is overlooked or not recorded because of its unapparent parasitism (as with *Rhamphicarpa*

 Parasitic Plants in African Agriculture (L.J. Musselman and J. Rodenburg)
DOI: 10.1079/9781789247657.0006

fistulosa), remote locations or a relative shortage in frequent and systematic weed surveys on the continent remains an open question.

6.2 Taxonomy

In their vegetative stage, plants of *Buchnera hispida* resemble *Striga* spp. morphologically, due to their stature and the shape of their leaves (Fig. 6.1). After completion of their life cycle, dead plants of this species dry black, which is another feature it shares with witchweeds. It is probably for these reasons, as well as overlapping host ranges, that *B. hispida* is sometimes confused with *Striga* spp. in local names and references (Nwoke and Okonkwo, 1974; J. Rodenburg, personal observation). Compared with *Striga* spp., however, it can be taller (about 1 m) with smaller flowers. The flowers are purple to blue, radially symmetrical, positioned in the axils of long bracts, and produce somewhat bigger seeds than those of *Striga* spp., at around 0.55 mm in length (Parker, 2013). Biologically it differs from *Striga* spp. because it is a facultative rather than an obligate parasite with germination and early growth being independent of the presence of a host plant.

6.3 Distribution

Buchnera is one of the few genera of Orobanchaceae hemiparasites to be represented throughout much of the world, especially in tropical regions. *Buchnera hispida* is distributed throughout sub-Saharan Africa, from Senegal to Ethiopia and from Sudan to South Africa, as well as Madagascar (Fig. 6.2).

Fig. 6.1. (A) Hairy buchnera (*Buchnera hispida*), observed in a field of sorghum, Ethiopia. (B) *B. hispida* on millet, Senegal.

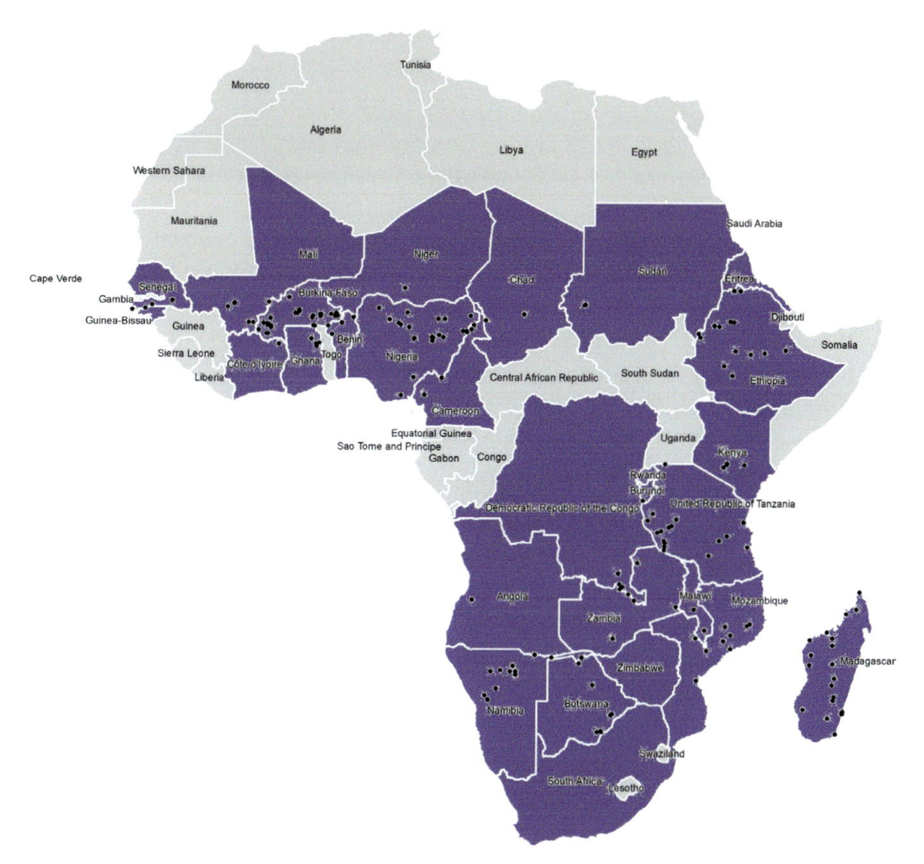

Fig. 6.2. *Buchnera hispida* distribution within Africa (mapped distribution depends on extent of collections). Data from Global Biodiversity Information Facility: GBIF.org (accessed 8 September 2022), GBIF Occurrence Download https://doi.org/10.15468/dl.gjgc8r. Black dots indicate georeferenced records from GBIF and non-grey colours indicate all countries where the species has been observed.

6.4 Hosts

The only extensive host range study in the genus found that a species native to the southern USA, *Buchnera americana*, formed attachments on 18 of 19 tree species including both gymnosperms and angiosperms (Musselman and Mann, 1978). In nature, *B. americana* forms haustoria on an equally diverse group of herbaceous plants as well. Whether the same breadth of hosts applies to *B. hispida* is unknown as the host range has not been extensively documented. However, the species seems restricted to tropical grasses, including cereal crops, as it is observed on common grasses such as Indian goosegrass (*Eleusine indica*) and Bermuda grass (*Cynodon dactylon*) (Parker and Riches, 1993) and cereals such as maize, sorghum and pearl millet (Nwoke and Okonkwo, 1974; Parkinson, 1989). It has also been observed in rice (Sallé *et al.*, 1994), but there are no other reports of this species as a problematic rice weed (Rodenburg *et al.*, 2010).

6.5 Biology and Ecology

As noted, *B. hispida* is a facultative parasitic plant. Similar to other facultative parasites, such as *R. fistulosa* (e.g. Ouédraogo *et al.*, 1999; Kabiri *et al.*, 2016), seed germination of *B. hispida* requires moisture and sunlight and is independent of host-plant-derived germination stimulants that drive seed germination in obligate parasitic plants (Okonkwo and Nwoke, 1974; Nwoke and Okonkwo, 1974, 1980). Nwoke and Okonkwo (1974) found it could be grown to maturity without sorghum but if a sorghum was in the pot, the parasite formed haustoria on the host. Microscopic sections of the connection confirmed a vascular connection between host and parasite. Like other facultative parasites, such as *Rhamphicarpa fistulosa* and *Rhinanthus minor* (Seel and Press, 1993; Kabiri *et al.*, 2016), parasitizing plants of *B. hispida* grow more vigorously (Nowke and Okonkwo, 1974). They grow taller with thicker stems and darker green leaves. Whether this also translates into a higher reproductive output, as with the other facultative parasitic plants noted, remains to be confirmed. While the parasite benefits from a host, parasitized host plants, in contrast, are greatly reduced in size compared with uninfected controls (Nwoke and Okonkwo, 1974).

Like *Striga* spp., *B. hispida* is found in free-draining soils (mostly sandy soils) and also seems to favour erratic rains and poor soil fertility conditions (Parker and Riches, 1993).

6.6 Control

There are no studies solely focusing on the control of *B. hispida*. It is, however, expected to be controlled in a similar way to other facultative parasitic weeds, such as *R. fistulosa*. Relatively low and concentrated infestations can be hand weeded or treated by post-emergence herbicides such as 2,4-dichlorophenoxyacetic acid. Some of the management strategies for *Striga* spp. are also likely to be effective against *B. hispida*, such as crop rotations with non-hosts. When *B. hispida* is observed, it is advised to prevent seed production; plants need to be removed or killed before or at early stages of flowering.

References

Gworgwor, N.A., Ndahi, W.B. and Weber, H.C. (2001) Parasitic weeds of north-eastern region of Nigeria: a new potential threat to crop production. In: *The BCP Conference, 13–15 November 2001, Brighton, UK*. British Crop Protection Council, Farnham, UK, pp. 181–186.

Kabiri, S., Van Ast, A., Rodenburg, J. and Bastiaans, L. (2016) Host influence on germination and reproduction of the facultative hemi-parasitic weed *Rhamphicarpa fistulosa*. *Annals of Applied Biology* 16, 144–154.

Musselman, L.J. and Mann, W.F. Jr (1978) *Root Parasites of Southern Forests. General Technical Report SO-20.* US Department of Agriculture, Forest Service, Washington, DC.

Nwoke, F.I.O. and Okonkwo, S.N.C. (1974) Facultative hemi-parasitism in *Buchnera hispida* Buch.-Ham. ex D. Don. *Annals of Botany* 38, 993–1002.

Nwoke, F.I.O. and Okonkwo, S.N.C. (1980) Photocontrol of seed-germination in the hemiparasite *Buchnera hispida* (Scrophulariaceae). *Physiologia Plantarum* 49, 388–392.
Okonkwo, S.N.C. and Nwoke, F.I.O. (1974) Seed germination in *Buchnera hispida* Buch.-Ham ex D. Don. *Annals of Botany* 38, 409–441.
Ouédraogo, O., Neumann, U., Raynal Roques, A., Sallé, G., Tuquet, C. *et al.* (1999) New insights concerning the ecology and the biology of *Rhamphicarpa fistulosa* (Scrophulariaceae). *Weed Research* 39, 159–169.
Parker, C. (2013) The parasitic weeds of the Orobanchaceae. In: Joel, D.M., Gressel, J. and Musselman, L.J. (eds) *Parasitic Orobanchaceae: Parasitic Mechanisms and Control Strategies.* Springer, Berlin, pp. 313–344.
Parker, C. and Riches, C.R. (1993) *Parasitic Weeds of the World: Biology and Control*. CAB International, Wallingford, UK.
Parkinson, V.O. (1989) A survey of the infestation of crops by *Striga* spp. in Benin, Nigeria and Togo. *Proceedings of the Nova Scotian Institute of Science* 39, 1–9.
Raynal-Roques, A. (1994) Major, minor and potential parasitic weeds in semi-arid tropical Africa: the example of Scrophulariaceae. In: Pieterse, A.H., Verkleij, J.A.C. and Ter Borg, S.J. (eds) *Proceedings of the Third International Workshop on Orobanche and Related Striga Research, 8–12 November 1993.* Royal Tropical Institute, Amsterdam, pp. 400–405.
Rodenburg, J., Riches, C.R. and Kayeke, J.M. (2010) Addressing current and future problems of parasitic weeds in rice. *Crop Protection* 29, 210–221.
Sallé, G., Raynal-Roques, A., Tuquet, C., Andary, C., Dembele, B. *et al.* (1994) *Striga* research in Mali, Burkina Faso and Senegal, supported by basic studies in France, in the framework of the STD2 project of the European Union. In: Pieterse, A.H., Verkleij, J.A.C. and Ter Borg, S.J. (eds) *Proceedings of the Third International Workshop on Orobanche and Related Striga Research, 8–12 November 1993.* Royal Tropical Institute, Amsterdam, pp. 700–709.
Seel, W.E. and Press, M.C. (1993) Influence of the host on three sub-Arctic annual facultative root hemiparasites. I. Growth, mineral accumulation and above-ground dry-matter partitioning. *New Phytologist* 125, 131–138.

7 Witchweed

Abstract

Among the 28 species of the obligate hemiparasitic witchweeds (genus *Striga*) in Africa only a few are known as weeds in food crops. Most of these weedy witchweed species present very serious production constraints in smallholder farming systems. The most important are *Striga hermonthica* and *S. asiatica* in cereal crops, primarily maize, millet rice and sorghum, and *S. gesnerioides* in legume crops, primarily cowpea. Other species briefly discussed in this chapter are *S. aspera*, *S. forbesii*, *S. passargei* and *S. brachycalyx*. *Striga hermonthica* is the dominant species in the Northern Hemisphere, whereas weedy forms of *S. asiatica* dominate below the Equator. *Striga gesnerioides* is problematic in Western Africa north of the Equator but also in Eastern and Southern Africa. These species have some common biological and agronomic features, including prolific seed production, small seed sizes, host plant dependency for seed germination and severe negative effects on host plant physiology, growth, development and production. Because of these characteristics and the fact that severe host damage occurs during the first 4–5 weeks after attachment, when the parasites are still underground and hence invisible, they are difficult to manage and cause significant yield losses. Witchweeds (mainly *S. asiatica* and *S. hermonthica*) are estimated to cause annual cereal production losses of at least 15 million t, equating to US$4.7 billion. These weeds can be partially controlled by the use of resistant crop varieties and cultural practices, such as intercropping and crop rotations with leguminous crops (for *S. asiatica* or *S. hermonthica*) or cereal crops (for *S. gesnerioides*), and preferably by an integrated management approach for more effective and long-term results. Feasibility of the implementation of an integrated approach for smallholder farmers in Africa is however questionable.

7.1 Introduction

The genus *Striga* Lour. includes some of the most widespread and problematic weed species in Africa. The origin of the name has been attributed to a witch because of the Italian name for witch, *strega*, derived from the Latin. Rather,

 Parasitic Plants in African Agriculture (L.J. Musselman and J. Rodenburg)
DOI: 10.1079/9781789247657.0007

Striga refers to the bristle-like hairs that are found in all the species, hairs that are botanically termed strigose. But it is appropriate they are known as witch-weeds, as the parasite severely affects its host during the belowground stages of its life cycle when it is still invisible, and once aboveground it produces 'bewitching', bright-coloured flowers (Fig. 7.1) at the expense of its host. It is because of these traits that smallholder farmers in Africa may not immediately make the connection between the flowering plants in their fields and the low yields, or even yield failures, of their crops caused by the witchweed.

7.2 The 'Big Three'

There are 28 species and six subspecies of witchweeds in Africa (Mohamed *et al*., 2001) but only a few species are known to be weedy in field crops. The most important are *S. asiatica* and *S. gesnerioides*, and perhaps the most serious of all in terms of human impact is *S. hermonthica*. *Striga asiatica* and *S. hermonthica* have similar host ranges, consisting of grasses, with cereals and sugarcane as the most important. *Striga gesnerioides* is the only weedy *Striga* species known to parasitize dicotyledonous plants.

Striga gesnerioides is an important weed in cowpea (e.g. Kamara *et al*., 2014; Horn *et al*., 2015), and to a lesser extent in sweet potato (Parker, 2012). It

Fig. 7.1. The agriculturally most important species of the genus *Striga*: (A) *Striga gesnerioides*; (B) *S. hermonthica*; (C) *S. asiatica*; (D) *S. aspera*.

is a well-documented parasite of tobacco, a relative of tomato. It is a potential threat to tomatoes, although to our knowledge this has never been investigated and we know of only one report of *S. gesnerioides* parasitizing tomatoes, which was in Zimbabwe (Knepper, 1989).

Striga hermonthica, *S. asiatica* and to lesser extent *S. aspera* are important weeds in cereal crops such as sorghum, maize (e.g. Kamara *et al.*, 2014; Lobulu *et al.*, 2019), pearl millet (e.g. Gigou *et al.*, 2009; Drabo *et al.*, 2019) and rice (Rodenburg *et al.*, 2010). *Striga forbesii* and *S. passargei* are lesser-known witchweeds but host damage has been reported and, unfortunately, these unfamiliar species have the potential to become serious problems. Farmers and agricultural workers should be aware of the possibility of increased spread and pathogenicity. For that reason, we have included some that have demonstrated crop damage as well as others that have potential to cause damage.

Despite decennia of research and agricultural technology and knowledge-transfer activities focused on the control of these parasitic weeds, they continue to impact crops and farmers' livelihoods in Africa (e.g. Ayongwa *et al.*, 2010; Scheiterle *et al.*, 2019). There are numerous underlying reasons for this (see Chapter 12, this volume).

The presence of witchweed is often linked with unfavourable environmental conditions, such as low soil fertility (Ayongwa *et al.*, 2010; Kamara *et al.*, 2014), sandy soils, areas of erratic rainfall or intensified, mono-cropping production systems (e.g. Cardwell and Lane, 1995; Dugje *et al.*, 2006; Dossou-Aminon *et al.*, 2016). Perhaps this is because under such harsh cropping circumstances, the witchweed-inflicted damage is likely to be more striking. However, *Striga* spp. are distributed throughout soil texture types, fertility levels and agro-climatic zones.

7.3 Effects on Host Plants

An early symptom of witchweed infection is the formation of brownish spots on host plant leaves (Fig. 7.2A,C). This leads to stunted growth, leaf rolling and senescence (Fig. 7.2B), along with reduced aboveground biomass and seed production of the host plant (Fig. 7.3). The biomass of the parasitic plants is often much smaller than the biomass losses incurred by the host plant (e.g. Graves *et al.*, 1989); this is often referred to as the pathological effect of *Striga* spp. This can be explained by parasite effects on host plant biochemistry, including decreased levels of the host plant growth regulators, cytokinins and gibberellic acid, and increasing levels of abscisic acid, leading to changes in host plant physiology including reduced stomatal conductance and photoinhibition resulting in impaired photosynthesis (e.g. Press *et al.*, 1987; Gurney *et al.*, 1995, 2002).

7.4 Impact

Witchweed-induced yield losses vary across *Striga* species and crop species, cultivars, soil fertility and agricultural input levels. Uncontrolled infection by *S. asiatica* can result in average yield losses of 73% in rice (Fig. 7.3; Rodenburg *et al.*, 2016) and 80% in maize (Ransom *et al.*, 1990). *Striga hermonthica*

Fig. 7.2. (A) Sorghum variety CK60B, with leaf spots as an early symptom of witchweed infection. (B) Stunted growth, leaf senescence and rolling as late symptoms of witchweed infection. (C) CK60B infected by *Striga hermonthica* in a variety screening trial in Mali.

decreased sorghum yields by 37% on average (ranging from 24% to 49% annually), across a range of sorghum genotypes with varying levels of resistance and tolerance. Almost no yield losses were observed for the genotype IS9830, which combined good resistance and tolerance, but there was 70% average yield loss for CK60B, which was susceptible and sensitive (Fig. 7.2; Rodenburg *et al.*, 2005). *Striga gesnerioides*-induced yield losses in cowpea are reported to frequently reach 30% (Aggarwal and Ouédraogo, 1989; Muleba *et al.*, 1997), but complete crop failure is not uncommon (Singh and Emechebe, 1990).

Fig. 7.3. Rice plants of variety IAC165 moderately (second and third pot from left to right) and heavily (last two pots) infected by *Striga asiatica* (red flowers) and *S. hermonthica* (purple flowers); note the reduced host plant height, tiller and panicle numbers in infected plants compared with witchweed-free plants (first and fourth pot).

Fig. 7.4. Millet crops in Côte d'Ivoire (A) and Mali (B), severely infected by *Striga hermonthica*, known as purple witchweed.

The economic impact of witchweed in Africa is not known. Based on the situation in six African countries, the total cropped area infested by *Striga* spp. in Africa was estimated at 21 million ha more than 30 years ago by Sauerborn (1991). Taking these estimates as a starting point and assuming an increase in distribution of 1.7% annually (Rodenburg *et al.*, 2016), the current arable crop area in Africa infested by *Striga* spp. would be estimated at 35.4 million ha in 2022, which equates to 38% of the area under the four main cereals, maize, sorghum, millet and rice. How this is distributed over the different cereal crops cannot be ascertained, but the most recent estimate of the rice area infested by *Striga* spp. is only 12% (Rodenburg *et al.*, 2016).

Rodenburg *et al.* (2016) estimated the annual upland rice production losses due to witchweed at 293,000 t. Based on this estimate, the proportionate loss, which is the total witchweed-inflicted production loss in Africa divided by the total estimated production of upland rice in Africa, is estimated at 12%. Applying this same proportionate witchweed-inflicted loss of 12% across other affected cereal crops (maize, millet and sorghum), using actual estimates of areas, yields and farm-gate prices from FAO (2022), the current estimated annual witchweed-inflicted cereal production losses in Africa amount to over 15 million t of grain, worth US$4.7 billion (Table 7.1).

On African smallholder farms, witchweed infestation implies not only a direct cost associated with yield losses but also an indirect cost, for example by using production resources to address infestation and by taking infested fields out of production. Most smallholders remove witchweed by hand or using a hand-held hoe (Mrema *et al.*, 2017; Tippe *et al.*, 2017a) and this requires large labour inputs leading to inefficiency (N'cho *et al.*, 2019).

7.5 Taxonomy

Striga gesnerioides (Willd.) Vatke is a branched, usually short (<30 cm) plant with small, scale leaves and low chlorophyll content. The colour of the flowers ranges from white, yellow, pink to purple (Fig. 7.1A). A more detailed description can be found in Parker (2013) and Mohamed *et al.* (2001).

Striga hermonthica (Del.) Benth. is the largest of the weedy *Striga* species. It is known by the English common names purple witchweed and giant witchweed. It is erect, often branched around 50 cm tall when mature, but plants can be up to 1 m. The ecotypes of Eastern Africa are usually taller than those from Western Africa. Stems and leaves have a rough surface texture. The leaves are narrow, oval shaped, about 3–8 cm in length and up to 1 cm in width. Flower colour is pale to dark pink (Figs 7.1B, 7.4), although occasionally white corolla mutants are observed. On very rare occasions albino plants, totally lacking chlorophyll, are found. An account of albino *S. hermonthica* is documented in Musselman and Hepper (1986). This albinism deserves more attention. In non-parasitic plants it would be a lethal trait. This is the only report of an achlorophyllous condition in a facultative root parasite that we are aware of. Further research into the genomes of albino versus chlorophyllous *S. hermonthica* could provide meaningful insight into the evolution of holoparasitism. A more detailed taxonomic description can be found in Parker (Parker, 2013) and Mohamed *et al.* (2001).

Table 7.1. Best-bet (conservative) estimates of total cropped area for each cereal crop, the area infested by witchweed, and the total current production and economic losses caused by witchweed.

Crop	Total cropped area (1,000 ha)	Estimated witchweed-infested crop area (1,000 ha)	Total production losses (1,000 t)	Total economic losses (US$ 1,000,000)
Rice (upland)	3,231	388	293	158
Maize	43,042	NA[a]	8,880	2,351
Millet	19,723	NA	2,208	797
Sorghum	27,291	NA	3,712	1,360
Total	**93,287**	**35,400**	**15,093**	**4,666**

[a]Reliable estimates of infested areas for maize, sorghum and millet are not available (NA).

Striga asiatica (L.) Kuntze (syn. *S. lutea* Lour.), commonly known as red witchweed or Asiatic witchweed, has the same general host range as *S. hermonthica* but the plants are much smaller and have distinctly different flowers, both in shape and colour (Fig. 7.1C). Although the colour of the most common ecotype is mainly red, yellow ecotypes are frequently observed. A more detailed description can be found in Parker (Parker, 2013) and Mohamed *et al.* (2001). The effect on the host appears similar for both *S. asiatica* and *S. hermonthica* (Fig. 7.3).

Striga aspera (Willd.) Benth. resembles *S. hermonthica* in morphology and flower colour and is often confused with it but is usually smaller and less robust. The chief difference between the species is the position of the bend in the corolla tube. The bend in the corolla tube in *S. hermonthica* is just above the calyx, whereas in *S. aspera* it is close to the corolla lobes (Fig. 7.5). *Striga aspera* also shares some morphological similarity with *S. brachycalyx*.

7.6 Distribution

Striga species are widely distributed in Africa (Figs 7.6 and 7.7). The most widespread species is *Striga asiatica,* which is found in at least 44 countries, while *S. hermonthica* is observed in 32 (Rodenburg *et al.*, 2016) and *S. gesnerioides* in 31 countries (Mohamed *et al.*, 2001). Among the four most important weedy *Striga* species, *S. aspera* has the most restricted distribution with reports from 17 countries (Rodenburg *et al.*, 2016). *Striga hermonthica* is most frequent north of the Equator, whereas *S. asiatica,* in particular the red-coloured weedy ecotype of this species, is mostly south of the Equator. The latter is also the only *Striga* species observed to be a weed problem in cereals in Madagascar

Fig. 7.5. The resemblance and differences (highlighted by white circles) between *Striga aspera* and *S. hermonthica*, Côte d'Ivoire.

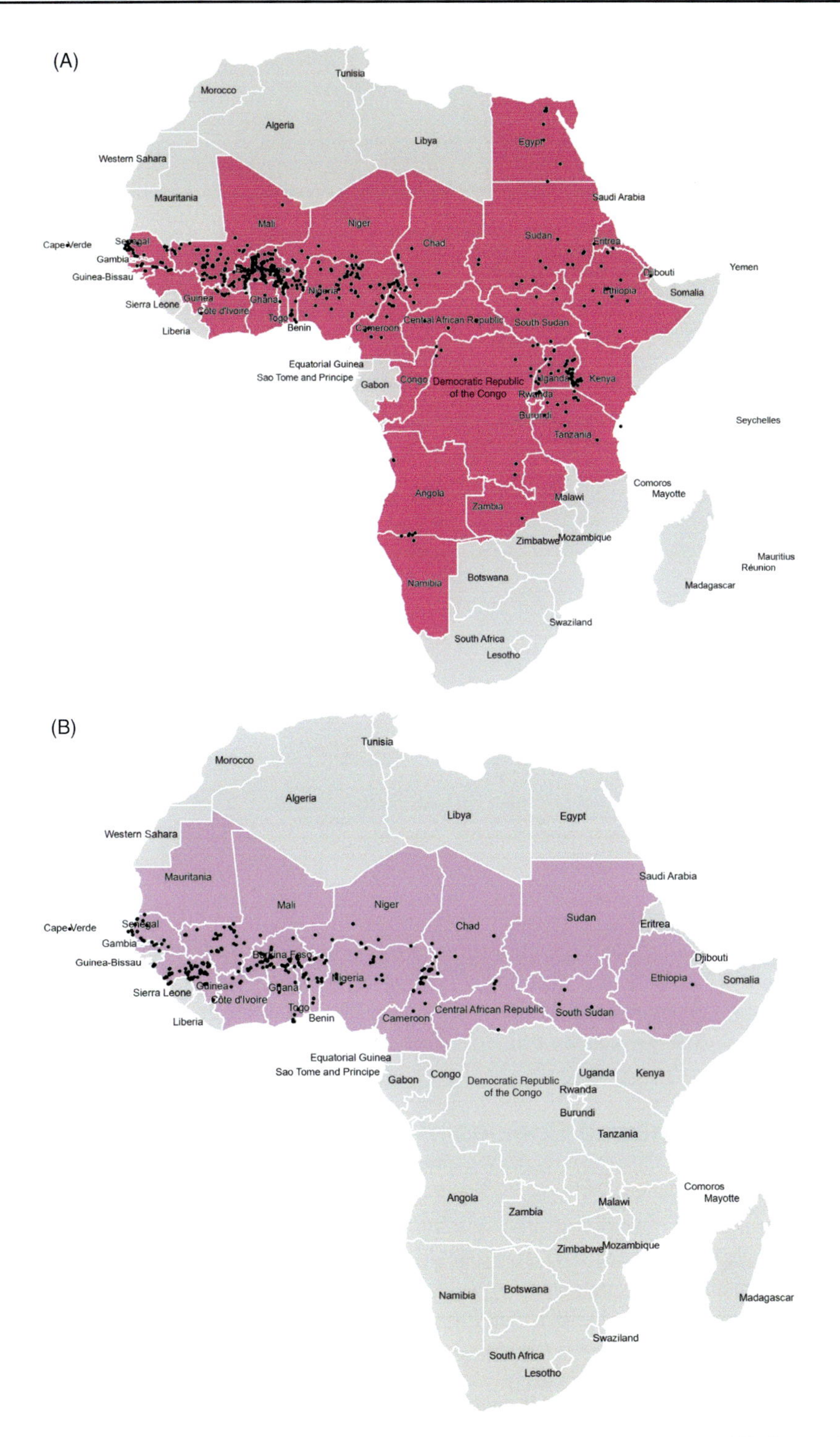

Fig. 7.6. Distribution of *Striga* spp. in Africa: (A) *S. hermonthica*; (B) *S. aspera*; (C) *S. asiatica*. Non-grey colours indicate all countries where the species has been observed. Adapted from Rodenburg *et al*. (2016).

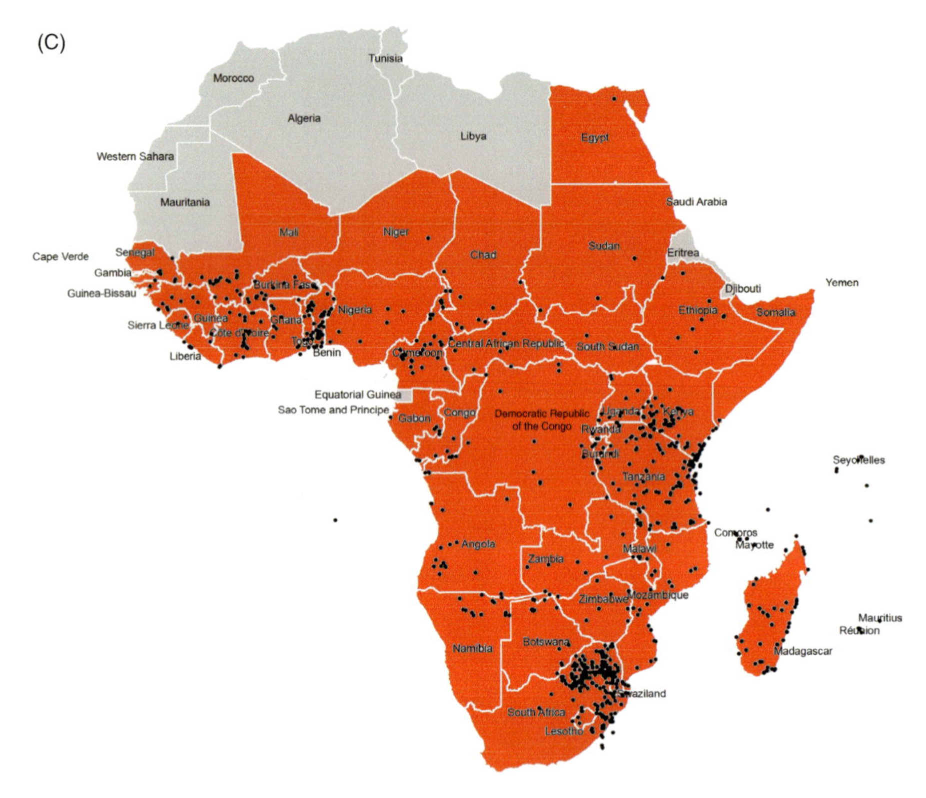

Fig. 7.6. Continued.

(Fig. 7.6C), while *S. gesnerioides* is the only *Striga* species found in Morocco (Fig. 7.7). There are numerous farmer surveys indicating the crop area under *Striga* spp. infestation is increasing over time (e.g. Dugje *et al.*, 2006; Aflakpui *et al.*, 2008; Kouakou *et al.*, 2015).

7.7 Other Witchweed Species of Agronomic Importance

Although the 'Big Three' (*S. asiatica*, *S. hermonthica* and *S. gesnerioides*) are the most widely distributed witchweeds and therefore cause the most damage to crops, there are two additional species of witchweed recorded as harming crops. Although less frequently encountered, these species need to be recognized as possible pests in the future. These are *Striga passargei* Engl. (no English common name) and *S. forbesii* Benth. (common name, giant maize witchweed).

Striga passargei is largely restricted to Western Africa (Fig. 7.8). It has been reported to be of limited impact in Burkina Faso and along the border in Mali, where it has been described as severe but localized on sorghum (Fig. 7.9; Raynal-Roques, 1994).

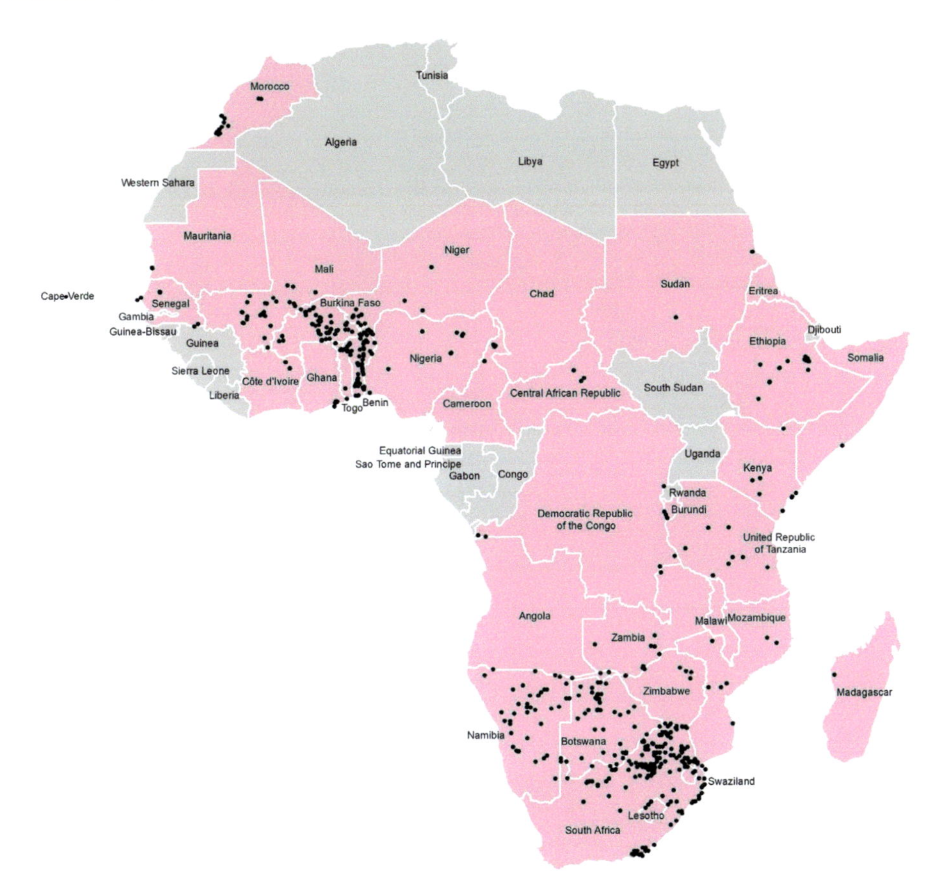

Fig. 7.7. *Striga gesnerioides* distribution in Africa (mapped distribution depends on extent of collections). Data from Global Biodiversity Information Facility: GBIF.org (accessed 10 September 2022), GBIF Occurrence Download https://doi.org/10.15468/dl.j2r6pe. Dots indicate georeferenced records from GBIF and non-grey colours indicate all countries where the species has been observed.

More serious is *S. forbesii*, giant maize witchweed (an unfortunate common name not to be confused with giant witchweed, *S. hermonthica*), which has been a serious impediment to the growth of maize and sorghum in Zimbabwe (Knepper, 1989). Sporadically, it has caused damage in Western African sorghum and maize (Knepper, 1989). It is widespread in sub-Saharan and Southern Africa where it also attacks sugarcane (Fig. 7.10). *Striga forbesii* was a major constraint in the Juba River Sugar Project in extreme southern Somalia (L.J. Musselman, unpublished). *Striga aspera* and *S. asiatica* also attack sugarcane. The impact of these small herbaceous parasites on the large, woody sugarcane is striking.

7.8 Witchweeds to Watch

There is a diversity of witchweeds in Africa, and it is not known why some are severe parasites impacting the yield and quality of host grains and legumes and

Fig. 7.8. *S. brachycalyx*, *S. forbesii* and *S. passargei* distribution in Africa (mapped distribution depends on extent of collections). Data from GBIF.org (accessed 8 September 2022) GBIF Occurrence Download https://doi.org/10.15468/dl.t9cysz, https://doi.org/10.15468/dl.pkqeuw and https://doi.org/10.15468/dl.4p26qp. Black dots indicate georeferenced records from GBIF and non-grey colours indicate all countries where the species has been observed.

others are restricted to native grasslands. An ill-defined third group of *Striga* species that have not yet proven to be a problem but have the potential to be so, especially as their habitats are invaded by expanding agriculture, has *S. brachycalyx* as its most prominent member.

Raynal-Roques (1994) has suggested criteria for identifying potential parasitic weeds: (i) the suspects will be related to species of witchweeds currently causing serious problems and therefore could be pre-adapted to becoming a problem; she cites the putatively close relationship between *Striga brachycalyx* (restricted to sub-Saharan Western Africa; Figs 7.8 and 7.11) and *S. aspera* (see above); (ii) suspects, *S. brachycalyx* for example, grow on both annual and perennial grasses allowing a longer growing season for the parasite and more seed production; (iii) suspects can grow in different moisture regimes; and (iv) a witchweed deserves special attention as a suspect if the species attacks relatives of food crops.

Fig. 7.9. (A) *Striga passargei* on sorghum, Tambacounda, Senegal. (B) Variation in corolla colour from a single population. (C) *S. passargei* on sorghum in a pot experiment at the Weed Research Organization (L.J. Musselman, unpublished).

7.9 Biology

Species of the genus are all obligate hemiparasites. This implies that they are fully dependent on a host plant during the belowground stages of their development. Seed germination of *Striga* spp. requires a period of seed preconditioning, under moist and warm conditions (25°C–30°C), and subsequently exposure to compounds released by host-plant roots (Bouwmeester *et al.*, 2003). There is a range of organic compounds that stimulate seed germination in *Striga* spp., but the most effective ones are found in a class of plant-growth regulators known as strigolactones (Cardoso *et al.*, 2011). After germination, *Striga* spp. radicles grow chemotropically towards a host root. If the *Striga* spp. radicle is then exposed to haustorium-inducing factors, a diverse group of generic organic compounds indicating contact with a host root (Yoshida

Fig. 7.10. (A) *Striga forbesii* attacking sorghum near Bulawayo, Zimbabwe. (B) *S. forbesii* on sugarcane, Juba River, Somalia.

Fig. 7.11. *S. brachycalyx* on sugarcane in Burkina Faso. Although a distinct species, this witchweed has flowers like those of *S. aspera.*

et al., 2016; Cui *et al.*, 2018; Goyet *et al.*, 2019), it starts the generation of a haustorium. Witchweeds that have been studied connect with their host through a terminal haustoria, a unique morphogenesis of the root tip developing directly into a haustorium (Masumoto *et al.*, 2021). Some witchweed species, such as *S. hermonthica*, are strictly cross-pollinated but others, such as *S. asiatica* and *S. gesnerioides*, are self-pollinated (autogamous). Witchweeds are known to produce copious amounts of seed, with estimates ranging from 6700 to 85,000 seeds per plant, depending on witchweed ecotype and host-plant genotype (Stewart, 1990; Webb and Smith, 1996; Rodenburg *et al.*, 2006). Witchweed plants continue to produce seeds beyond crop harvest (Rodenburg *et al.*, 2006).

7.10 Comparison of *Striga* Spp. with Other Parasitic Weeds

Alectra vogelii Benth. has a host range like that of *Striga gesnerioides* and is common in cowpea production systems alongside *S. gesnerioides* (e.g. Horn *et al.*, 2015). In this respect the cowpea lines with resistance against both parasitic weeds are a valuable finding for smallholder farmers. *Rhamphicarpa fistulosa* (Hochst.) Benth., despite a distinctly different appearance, is occasionally mistaken for *Striga* spp. Farmers in Uganda have called it the 'Striga of rice' (see Section 5.8, this volume).

7.11 Control

The control of witchweed is technically attainable, as shown by successful *S. asiatica* eradication efforts in the USA starting in the 1960s (Eplee, 1992). This was achieved by a very intensive and costly programme of detection, regulation and systematic control using pre- and post-emergence herbicides and ethylene (C_2H_4) gas injections in witchweed-infested soils, combined with strict quarantine measures. In sub-Saharan Africa, where witchweed is primarily a problem in smallholder farms, such capital and input-intensive approaches are not feasible. Solutions that are feasible for smallholder farmers in Africa should be effective but above all accessible and affordable (Tippe *et al.*, 2017a).

A wide range of technologies have been developed in the past decades, including resistant and tolerant crop varieties (e.g. Singh and Emechebe, 1990; Rodenburg *et al.*, 2005; Menkir and Kling, 2007; Rodenburg *et al.*, 2017), changes in crop establishment timing (Ekeleme *et al.*, 2011; Tippe *et al.*, 2017b), the use of cover crops such as the push–pull technology (e.g. Khan *et al.*, 2010; Pickett *et al.*, 2010) and conservation agriculture practices (Randrianjafizanaka *et al.*, 2018). Delaying parasite infection contributes to reduced effects on the host, information that can be used for the design of an effective management strategy (Gurney *et al.*, 1999; Van Ast and Bastiaans, 2006). Integration of some of these strategies has been proven to provide the most effective and probably most durable forms of control (Tesso and Ejeta, 2011; Tippe *et al.*, 2017a; Kanampiu *et al.*, 2018), but the feasibility of integrated approaches for smallholder farmers remains to be investigated. More detail on the control of witchweeds is provided in Chapter 13, this volume.

References

Aflakpui, G.K.S., Bolfrey-Arku, G.E.K., Anchirinah, V.M., Manu-Aduening, J.A. and Adu-Tutu, K.O. (2008) Incidence and severity of *Striga* spp. in the coastal savanna zone of Ghana: results and implications of a formal survey. *Outlook on Agriculture* 37, 219–224.

Aggarwal, V.D. and Ouédraogo, J.T. (1989) Estimation of cowpea yield loss from *Striga* infestation. *Tropical Agriculture* 66, 9192.

Ayongwa, G.C., Stomph, T.J., Hoevers, R., Ngoumou, T.N. and Kuyper, T.W. (2010) *Striga* infestation in northern Cameroon: magnitude, dynamics and implications for management. *NJAS – Wageningen Journal of Life Sciences* 57, 159–165.

Bouwmeester, H.J., Matusova, R., Sun, Z.K. and Beale, M.H. (2003) Secondary metabolite signalling in host–parasitic plant interactions. *Current Opinion in Plant Biology* 6, 358–364.

Cardoso, C., Ruyter-Spira, C. and Bouwmeester, H.J. (2011) Strigolactones and root infestation by plant-parasitic *Striga*, *Orobanche* and *Phelipanche* spp. *Plant Science* 180, 414–420.

Cardwell, K.F. and Lane, J.A. (1995) Effect of soils, cropping system and host phenotype on incidence and severity of *Striga gesnerioides* on cowpea in West Africa. *Agriculture, Ecosystems & Environment* 53, 253–262.

Cui, S., Wada, S., Tobimatsu, Y., Takeda, Y., Saucet, S.B. *et al.* (2018) Host lignin composition affects haustorium induction in the parasitic plants *Phtheirospermum japonicum* and *Striga hermonthica*. *New Phytologist* 218, 710–723.

Dossou-Aminon, I., Dansi, A., Ahissou, H., Cisse, N., Vodouh, R. *et al.* (2016) Climate variability and status of the production and diversity of sorghum (*Sorghum bicolor* (L.) Moench) in the arid zone of northwest Benin. *Genetic Resources and Crop Evolution* 63, 1181–1201.

Drabo, I., Zangre, R.G., Danquah, E.Y., Ofori, K., Witcombe, J.R. *et al.* (2019) Identifying farmers' preferences and constraints to pearl millet production in the Sahel and North-Sudan zones of Burkina Faso. *Experimental Agriculture* 55, 765–775.

Dugje, I.Y., Kamara, A.Y. and Omoigui, L.O. (2006) Infestation of crop fields by *Striga* species in the savanna zones of Northeast Nigeria. *Agriculture, Ecosystems & Environment* 116, 251–254.

Ekeleme, F., Kamara, A.Y., Omoigui, L.O., Chikoye, D., Dugje, I.Y. *et al.* (2011) Effect of sowing date on *Striga* infestation and yield of sorghum (*Sorghum bicolor* (L.) Moench) cultivars in the Sudan savanna of Northeast Nigeria. *African Journal of Agricultural Research* 6, 3240–3246.

Eplee, R.E. (1992) Witchweed (*Striga asiatica*): an overview of management strategies in the USA. *Crop Protection* 11, 3–7.

FAO (2022) FAOSTAT. http://faostat.fao.org/ (accessed 25 July 2022).

Gigou, J., Stilmant, D., Diallo, T.A., Cissé, N. and Sanogo, M.D. *et al.* (2009) Fonio millet (*Digitaria exilis*) response to N, P and K fertilizers under varying climatic conditions in West Africa. *Experimental Agriculture* 45, 401–415.

Goyet, V., Wada, S., Cui, S., Wakatake, T., Shirasu, K. *et al.* (2019) Haustorium inducing factors for parasitic Orobanchaceae. *Frontiers in Plant Science* 10: 1056.

Graves, J.D., Press, M.C. and Stewart, G.R. (1989) A carbon balance model of the sorghum–*Striga hermonthica* host–parasite association. *Plant, Cell & Environment* 12, 101–107.

Gurney, A.L., Press, M.C. and Ransom, J.K. (1995) The parasitic angiosperm *Striga hermonthica* can reduce photosynthesis of its sorghum and maize hosts in the field. *Journal of Experimental Botany* 46, 1817–1823.

Gurney, A.L., Press, M.C. and Scholes, J.D. (1999) Infection time and density influence the response of sorghum to the parasitic angiosperm *Striga hermonthica*. *New Phytologist* 143, 573–580.

Gurney, A.L., Taylor, A., Mbwaga, A., Scholes, J.D. and Press, M.C. (2002) Do maize cultivars demonstrate tolerance to the parasitic weed *Striga asiatica*? *Weed Research* 42, 299–306.

Horn, L., Shimelis, H. and Laing, M. (2015) Participatory appraisal of production constraints, preferred traits and farming system of cowpea in the northern Namibia: implications for breeding. *Legume Research* 38, 691–700.

Kamara, A.Y., Ekeleme, F., Jibrin, J.M., Tarawali G. and Tofa, I. (2014) Assessment of level, extent and factors influencing *Striga* infestation of cereals and cowpea in a Sudan savanna ecology of northern Nigeria. *Agriculture, Ecosystems & Environment* 188, 111–121.

Kanampiu, F., Makumbi, D., Mageto, E., Omanya, G., Waruingi, S. *et al.* (2018) Assessment of management options on *Striga* infestation and maize grain yield in Kenya. *Weed Science* 66, 516–524.

Khan, Z.R., Midega, C.O., Bruce, T.J.A., Hooper, A.M. and Pickett, J.A. (2010) Exploiting phytochemicals for developing a 'push–pull' crop protection strategy for cereal farmers in Africa. *Journal of Experimental Botany* 61, 4185–4196.

Knepper, D.A. (1989) Studies on the giant mealie witchweed, *Striga forbesii* Benth. in Zimbabwe. MSc thesis, Old Dominion University, Norfolk, Virginia.

Kouakou, K.C., Akanvou, L., Zoro Bi, I.A., Akanvou, R. and N'da, H.A. (2015) Distribution des espèces de *Striga* et infestation des cultures céréalières dans le nord de la Côte d'Ivoire. *Cahiers Agricultures* 24, 37–46.

Lobulu, J., Shimelis, H., Laing, M. and Mushongi, A.A. (2019) Maize production constraints, traits preference and current *Striga* control options in western Tanzania: farmers' consultation and implications for breeding. *Acta Agriculturae Scandinavica Section B - Soil and Plant Science* 69, 734–746.

Masumoto, N., Suzuki, Y., Cui, S., Wakazaki, M., Sato, M. *et al.* (2021) Three-dimensional reconstructions of haustoria in two parasitic plant species in the Orobanchaceae. *Plant Physiology* 185, 1429–1442.

Menkir, A. and Kling, J.G. (2007) Response to recurrent selection for resistance to *Striga hermonthica* (Del.) Benth. in a tropical maize population. *Crop Science* 47, 674–682.

Mohamed, K.I., Musselman, L.J. and Riches, C.R. (2001) The genus *Striga* (Scrophulariaceae) in Africa. *Annals of the Missouri Botanical Garden* 88, 60–103.

Mrema, E., Shimeli, H., Laing, M. and Bucheyeki, T. (2017) Farmers' perceptions of sorghum production constraints and *Striga* control practices in semi-arid areas of Tanzania. *International Journal of Pest Management* 63, 146–156.

Muleba, N., Ouédraogo, J.T. and Tignegre, J.B. (1997) Cowpea yield losses attributed to *Striga* infestations. *Journal of Agricultural Science* 129, 43–48.

Musselman, L.J. and Hepper, F.N. (1986) The witchweeds (Striga, Scrophulariaceae) of the Sudan Republic. *Kew Bulletin* 41, 105–221.

N'cho, S.A., Mourits, M., Rodenburg, J. and Lansink, A.O. (2019) Inefficiency of manual weeding in rainfed rice systems affected by parasitic weeds. *Agricultural Economics* 50, 151–163.

Parker, C. (2012) Parasitic weeds: a world challenge. *Weed Science* 60, 269–276.

Parker, C. (2013) The parasitic weeds of the Orobanchaceae. In: Joel D.M., Gressel, J. and Musselman, L.J. (eds) *Parasitic Orobanchaceae: Parasitic Mechanisms and Control Strategies*. Springer, Berlin, pp. 313–344.

Pickett J.A., Hamilton, M.L., Hooper, A.M., Khan, Z.R. *et al.* (2010) Companion cropping to manage parasitic plants. *Annual Review of Phytopathology* 48, 161–177.

Press, M.C., Tuohy, J.M., Stewart, G.R., (1987) Gas exchange characteristics of the sorghum Striga host–parasite association. *Plant Physiology* 84, 814–819.

Randrianjafizanaka, M.T., Autfray, P., Andrianaivo, A.P., Ramonta, I.R. and Rodenburg, J. (2018) Combined effects of cover crops, mulch, zero-tillage and resistant varieties on *Striga asiatica* (L.) Kuntze in rice–maize rotation systems. *Agriculture, Ecosystems & Environment* 256, 23–33.

Ransom, J.K., Eplee, R.E. and Langston, M.A. (1990) Genetic variability for resistance to *Striga asiatica* in maize. *Cereal Research Communications* 18, 329–334.

Raynal-Roques, A. (1994) Major, minor and potential parasitic weeds in semi-arid tropical Africa: the example of Scrophulariaceae. In: Pieterse, A.H., Verkleij, J.A.C. and Borg, S.J. (eds) *Biology and Management of Orobanche. Proceedings of the Third International Workshop on Orobanche and Related Striga Research*. Royal Tropical Institute, Amsterdam, pp. 400–405.

Rodenburg, J., Bastiaans, L., Weltzien, E. and Hess, D.E. (2005) How can field selection for *Striga* resistance and tolerance in sorghum be improved? *Field Crops Research* 9, 34–50.

Rodenburg, J., Bastiaans, L., Kropff, M.J. and Van Ast, A. (2006) Effects of host plant genotype and seedbank density on *Striga* reproduction. *Weed Research* 46, 251–263.

Rodenburg, J., Riches, C.R., and Kayeke, J.M. (2010) Addressing current and future problems of parasitic weeds in rice. *Crop Protection* 29, 210–221.

Rodenburg, J., Demont, M., Zwart, S.J. and Bastiaans, L. (2016) Parasitic weed incidence and related economic losses in rice in Africa. *Agriculture, Ecosystems & Environment* 235, 306–317.

Rodenburg, J., Cissoko, M., Kayongo, N., Dieng, I., Bisikwa, J. *et al.* (2017) Genetic variation and host–parasite specificity of *Striga* resistance and tolerance in rice: the need for predictive breeding. *New Phytologist* 21, 1267–1280.

Sauerborn, J. (1991) The economic importance of the phytoparasites Orobanche and Striga. In: Ransom, J.K., Musselman, L.J., Worsham, A.D. and Parker, C. (eds) *Proceedings of the 5th International Symposium of Parasitic Weeds, Nairobi, Kenya, 24–30 June 1991*. CIMMYT, pp. 137–143.

Scheiterle, L., Haring, V., Birner, R. and Bosch, C. (2019) Soil, *Striga*, or subsidies? Determinants of maize productivity in northern Ghana. *Agricultural Economics* 50, 479–494.

Singh, B.B. and Emechebe, A.M. (1990) Inheritance of *Striga* resistance in cowpea genotype B301. *Crop Science* 30, 879–881.

Stewart, G. (1990) Witchweed: a parasitic weed of grain crops. *Outlook on Agriculture* 19, 115–117.

Tesso, T.T. and Ejeta, G. (2011) Integrating multiple control options enhances *Striga* management and sorghum yield on heavily infested soils. *Agronomy Journal* 103, 1464–1471.

Tippe, D.E., Rodenburg, J., Schut, M., Van Ast, A., Kayeke, J. *et al.* (2017a) Farmers' knowledge, use and preferences of parasitic weed management strategies in rain-fed rice production systems. *Crop Protection* 99, 93–107.

Tippe, D.E., Rodenburg, J., Van Ast, A., Anten, N.P.R., Dieng, I. *et al.* (2017b) Delayed or early sowing: timing as parasitic weed control strategy in rice is species and ecosystem dependent. *Field Crops Research* 214, 14–24.

Van Ast, A. and Bastiaans, L. (2006) The role of infection time in the differential response of sorghum cultivars to *Striga hermonthica* infection. *Weed Research* 46, 264–274.

Webb, M. and Smith, M.C. (1996) Biology of *Striga hermonthica* (Scrophulariaceae) in Sahelian Mali: effects on pearl millet yield and prospects of control. *Weed Research* 36, 203–311.

Yoshida, S., Cui, S., Ichihashi, Y. and Shirasu, K. (2016) The haustorium, a specialized invasive organ in parasitic plants. *Annual Review of Plant Biology* 67, 643–667.

8 Alectra

Abstract

Apart from *Striga* spp., the only other major obligate (root) hemiparasitic weed in Africa is *Alectra vogelii*. Although the genus *Alectra* contains around 30 species, only four species are known as weeds in African agriculture; apart from *A. vogelii*, these are *A. picta*, *A. orobanchoides* and *A. sessiliflora*. In African farming systems *Alectra vogelii* is the most impactful weed species of the genus. Because it is an obligate hemiparasite with similarity to *Striga* spp. in terms of biology and host interactions, and owing to its yellow flower colour, it is known as 'yellow witchweed'. The species has a similar distribution and (albeit broader) host range to *Striga gesnerioides* but is morphologically distinctly different as it is a more robust plant with a more pronounced foliage. As for *S. gesnerioides*, the most frequent host crop species is cowpea, but *A. vogelii* can also be found in groundnut, Bambara nut, mung bean, lablab, common bean and occasionally chickpea and soybean. *Alectra vogelii* and the related *A. picta* can be controlled in much the same way as *S. gesnerioides*. A number of cowpea varieties with resistance against *S. gesnerioides* also proved resistant against *A. vogelii*. In addition to the use of resistant crop varieties, crop rotations with adapted non-host species like millet, sorghum, pigeon pea or sunflower, as trap crops, could be an effective management approach. Other control practices are discussed as well.

8.1 Introduction

Alectra is a genus of about 30 species (Smith *et al*., 2001), mostly African but with some South American representatives. No standard common names are widely used. The most wide-ranging species, *A. vogelii*, is referred to as 'yellow witchweed'. There are some species, or ecotypes of species, of witchweed (*Striga* spp.) with yellow flowers (e.g. *S. asiatica* in Western Africa) but the flowers of *Alectra* spp. bear little resemblance to those of *Striga* spp. so it is preferable to refer to them as alectra.

 Parasitic Plants in African Agriculture (L.J. Musselman and J. Rodenburg)
DOI: 10.1079/9781789247657.0008

Alectra spp. are herbaceous hemiparasites with opposite, hairy leaves and usually fleshy basal stems. The flowers are rotate (radially symmetrical) with five petals and an extended style (Figs 8.1, 8.2, 8.3). Like their relatives in the genus *Striga*, *Alectra* spp. produce capsules with copious tiny seeds.

So far only five species have been reported to damage crops, of which four can be found in Africa. Descriptions and identification keys are presented in

Fig. 8.1. (A) *Alectra sessiliflora* showing the characteristic rotate corolla. (B) *A. vogelii* on groundnut in Mali. The corolla shape and colour are similar in these two species. In general, the leaves of the first species are larger.

Fig. 8.2. (A) *Alectra vogelii* on cowpea in Embu, central Kenya. Image used courtesy of H.J. Bouwmeester. (B) *A. vogelii* on cowpea in northern Namibia.

Fig. 8.3. *Alectra orobanchoides* parasitizing a shrubby member of the sesame family, *Sesamothamnus guerichii*, northern Namibia (A) Emerging stems; the red is characteristic of many plants in this species. (B) Parasite attached to the root of the host; the numerous buds arising at the point of attachment are evident.

Parker and Riches (1993). The most important are *A. vogelii* and *A. picta*, both of which can be a serious problem in legume crops (Sauerborn *et al.*, 2007).

8.2 Distribution

Alectra vogelii is found across semi-arid parts of Africa (Fig. 8.4; Parker and Riches, 1993; Parker, 2013) and is generally distributed in cowpea-growing areas. *Alectra vogelii* has been reported as a crop parasite in Burkina Faso in Western Africa (Dieni *et al.*, 2018), Ethiopia (Hussien *et al.*, 2006), Malawi (Phiri *et al.*, 2019), Botswana (Riches, 1989a) and Namibia (Horn *et al.*, 2015). *Alectra vogelii* has a similar host plant range to *Striga gesnerioides*. These parasites are often found together on farms in Namibia (Horn *et al.*, 2015). Similar cohorts are reported from Nigeria (Lagoke, 1989). Apart from cowpea, the most frequent host, *A. vogelii* may attack groundnut, Bambara nut, mung bean, lablab and common bean, and has occasionally been observed on chickpea and soybean. Although *A. vogelii* parasitizes most leguminous crops, pigeon pea seems to be the exception among the common pulse crops (Parker and Riches, 1993). There is one report of a non-legume crop, flax, hosting *A. vogelii* (Phiri *et al.*, 2019). Weeds parasitized by *A. vogelii* include the legumes *Tephrosia purpurea* and *Indigofera daleoides*, as well as non-leguminous species such as *Sesamum grandiflorum, Acanthospermum australe, A. hispidum, Acrotome inflata, Euphorbia chamaesyce, E. inaequilatera, Hibiscus* spp. and *Vernonia poskeana* (Riches, 1989a). Like *Striga gesnerioides, A. vogelii* has different location-specific strains (ecotypes). Some of these strains are unable to attack mungbean in Western

Fig. 8.4. Distribution of *Alectra vogelii* and *A. picta* species within Africa (mapped distribution depends on extent of collections). Data from Global Biodiversity Information Facility: GBIF.org (accessed 25 August 2022), GBIF Occurrence Download https://doi.org/10.15468/dl.c3zrtd and https://doi.org/10.15468/dl.xr8wkw). Dots indicate georeferenced records from GBIF and non-grey colours indicate countries where the species has been observed.

Africa and Cameroon, and Bambara nut in Western Africa, Cameroon, Botswana and areas of South Africa (Polniaszek and Parker, 1987; Riches *et al.*, 1992). Only strains from Eastern and Southern Africa, including Kenya, Malawi and the province Mpumalanga in South Africa, parasitize the full range of common hosts, cowpea, groundnut, mung bean and Bambara nut. In addition, strains from different locations may have different levels of virulence across varieties within a host species. *Alectra*-susceptible cowpea lines from Southern Africa, for instance, have shown resistance against *A. vogelii* strains from Western Africa (Polniaszek *et al.*, 1991).

Alectra picta has a similar host range to *A. vogelii*, including groundnut, Bambara nut and cowpea (Musango *et al.*, 2022), but a much more restricted distribution (Fig. 8.4). Because of their shared hosts and resemblance and because they can be found in the same field, *A. picta* and *A. vogelii* are likely to be misidentified. Reports on *A. picta* are therefore not always reliable. Compared with *A. vogelii*, *A. picta* is somewhat taller but with shorter and more slender leaves. An obvious difference is the stamen filaments, which are glabrous for *A. vogelii* but hairy for *A. picta* (Parker and Riches, 1993). Further data on the host range of these two *Alectra* species are needed.

Less frequent is *Alectra orobanchoides*, which is distinct within the genus in lacking well-developed leaves and having a reduced amount of chlorophyll, sometimes so sparse that the plant is considered a holoparasite. Distribution of this species is mainly restricted to Eastern and Southern Africa (Fig. 8.5). It causes important losses in sunflower and tobacco. There is one record of it parasitizing sesame (Visser *et al.*, 1984). A recent report (Musselman *et al.*, 2020) documents the parasitism of *Sesamothamnus guerichii*, a relative of sesame, in Namibia.

The most widespread *Alectra* species is *A. sessiliflora*, extending across Africa (Fig. 8.5) into western Asia as far as China (Morawetz and Wolfe,

Fig. 8.5. Distribution of *Alectra orobanchoides* and *A. sessiliflora* in Africa (mapped distribution depends on extent of collections). Data from GBIF.org (accessed 25 August 2022) GBIF Occurrence Download https://doi.org/10.15468/dl.dx4bfk and https://doi.org/10.15468/dl.kkta64. Dots indicate georeferenced records from GBIF and non-grey colours indicate countries where the species has been observed.

2011). *Alectra sessiliflora* has been reported from Ethiopia attacking niger seed (*Guizotia abyssinica*) (Parker, 1988), but apart from that, is very rarely reported attacking crops (J.J. Morawetz, personal communication).

8.3 Biology and Host Interactions

Alectra vogelii can be self- or cross-pollinated, but it always requires insects (Riches *et al.*, 1989b). Like *Striga* spp., *A. vogelii* is a highly prolific seed producer. The species may produce 200–350 seed capsules per plant, with 2000–3000 seeds per capsule, resulting in up to 600,000 seeds per plant (Botha, 1946; Visser, 1978). The seed size of *A. vogelii* is 0.15–0.25 mm (Visser, 1978). Seed germination is triggered by host-derived organic compounds (e.g. alectrol and strigol) in the host plant rhizosphere (Müller *et al.*, 1992). Seeds do not need to be preconditioned, as dry seeds exposed to germination stimulants germinate when incubated at 28°C–30°C (Visser and Johnson, 1982; Riches, 1989b). However, the sensitivity of the seeds to germination stimulants is enhanced if seeds have a preconditioning phase under moist and warm conditions. The optimum preconditioning is achieved after 9 days at 30°C–35°C. Although seed germination has similar requirements to that of *Striga* spp., an important difference is that *Alectra* spp. lack dormancy so newly produced seed can immediately germinate (Riches, 1989b). After germination, the parasite needs a suitable host within 10 days to survive. The success rate of the parasite radicle to find a suitable host is greatly increased by chemotropism (Visser, 1978).

The parasite attaches to the host root by haustorial hairs enabling the haustorium to penetrate into the host root tissue to form xylem–xylem connections (Fig. 8.3B; Visser *et al.*, 1990). Alectra also establishes a parenchyma bridge between the host and parasite phloem, facilitating the transfer of organic compounds (Dörr *et al.*, 1979). Emergence of the parasite generally occurs 4 weeks after germination with flowering around 2 weeks later (Visser, 1978).

As with *Striga* spp., the photosynthetic ability of alectra plants is limited. A study by de la Harpe *et al.* (1979) demonstrated that the photosynthetic rate of *A. vogelii* is only 21% that of sunflower. Consequently, most of the metabolites for growth of alectra plants are derived from the host (Gouws *et al.*, 1980). The impact on the host plant is evident, even before the parasite emerges. Common symptoms are wilting, delayed development, reductions in shoot biomass (relative to increase in root-biomass), rhizobium nodulation, and flower and pod numbers (Mugabe, 1983; Alonge *et al.*, 2001; Rambakudzibga *et al.*, 2002; Kureh and Alabi, 2003). Yield losses are moderate (15%; e.g. Salako, 1984) to severe (50%; e.g. Beck, 1987; Mbwaga *et al.*, 2000), and may be 100% in certain areas, such as in Embu district of Kenya (Bagnall-Oakeley *et al.*, 1991). As with other root-parasitic weeds, yield losses are a function of seed bank density, environmental conditions (primarily: timing, distribution and quantity of rainfall), crop species, and variety, implementation and effectiveness of weed control measures. Apart from yield reductions of infected host plants, *A. vogelii* also reduces mineral concentrations (Samson and Kehinde, 2009) and the total soluble carbohydrate content (Alonge *et al.*, 2005) in the grains of their hosts.

8.4 Control

Alectra can be controlled in the same way as *S. gesnerioides* and other *Striga* spp. (see Chapter 13, this volume).

8.4.1 Prevention

Preventive measures are ideally based on good agricultural practices and as such not only prevent infection and damage by *Alectra* spp. but also generally contribute to healthier and more productive crops and soils. Initial efforts should prevent contamination of new fields as well as prevention of a build-up of the alectra seed bank in infested fields. Both objectives, also relevant for other parasitic weeds, may not be easy to attain (see Chapter 12, this volume) and require awareness, vigilance and persistence in the application of measures.

In order for alectra-free fields to remain parasite free, crop seed sources should not contain alectra seeds and should ideally come from alectra-free fields (Parker and Riches, 1993). The dominance in Africa of informal seed systems lacking phytosanitary checks and measures, the low awareness of parasitic weed risks among farmers, extension or crop protection services (in particular those new to the problem) and the small sizes of alectra seeds make it nearly impossible to prevent contamination through planting materials. This may be particularly difficult for groundnut, as the pods are growing in the soil and could easily be contaminated.

Second, wind and water movements are likely factors in the spread of alectra seeds from infested to uninfested fields, and such movements should be prevented. Windbreaks around a field, for instance provided by living hedges, can be helpful to prevent import of seeds from contaminated fields. Permanently covering soils of contaminated fields by living or dead biomass (i.e. cover crops or mulch) could prevent alectra seeds from being wind-blown or water-transported from such fields into parasite-free fields. Water runoff can further be managed by soil bunds or drainage ditches around alectra-infested and alectra-free crop fields.

Measures discussed below may contribute to reduction, but there is no evidence that any of these, in particular when applied in isolation, reduces or eliminates the seed bank over time.

8.4.2 Mechanical control

Removal (by hand) and destruction of any parasitic weed plants from an infested field is a useful measure preventing seed bank increases if the infestation level is still low enough to carry out such practice. Hand weeding *before* flowering is recommended, as vegetative uprooted parasitic plants will not produce seed. After crop harvesting, removal of any surviving alectra plants by hand or, better yet, through soil tillage is recommended as well to prevent the parasites from continuing parasitism of the roots of the harvested crop (Parker and Riches,

1993). Damage to the crop cannot be completely prevented by mechanical control, requiring additional solutions.

8.4.3 Genetic control

In infested fields, a first line of defence is using alectra-resistant cultivars, able to reduce the number of successful infections that result in reproductive parasitic plants. Very few, if any, groundnut varieties are known to be resistant against *A. vogelii* or *A. picta*. Groundnut resistance screening work would therefore be a useful future research endeavour. Soybean observations from Malawi strongly indicate that the varieties Ocepara-4 and Bossier combine useful levels of resistance with satisfactory yields under alectra-infested conditions (Kabambe *et al.*, 2008). Another soybean variety, TGX1681-3F, known to be an effective trap crop against *Striga hermonthica*, appeared to have good resistance to infection by *Alectra* spp. as well. A number of cowpea accessions have been screened for *A. vogelii* resistance, including some varieties that are already adopted by farmers (e.g. IT99K-573-2-, IT98K-205-8 and Komcalle; Dieni *et al.*, 2018). Some cowpea accessions are resistant against both *A. vogelii* and *Striga gesnerioides*: IT90K-76, IT90K-82-2 and IT97K-819-154 according to Singh (2002), and B301 according to Riches (1989a). Cowpea variety B359 has shown superior and broad-range resistance against *Alectra* spp. from different countries (Riches *et al.*, 1992; Mainjeni, 1999). The recent identification of resistance markers in bean and asparagus bean (*Vigna unguiculata* ssp. *sesquipedalis*) could lead to future marker-assisted selection for *A. vogelii* resistance breeding (Ugbaa *et al.*, 2021). Across parasitic weed-resistant varieties, none of the identified materials is immune to *Alectra* spp., as with resistance against witchweeds.

8.4.4 Cultural control

The use of non-host species as trap crops in rotation with alectra host species, would be an effective solution meeting the above requirements; crop rotations can break crop-specific pest and disease cycles and may contribute to more effective soil nutrient cycling and use efficiency. Millet, sorghum or pigeon pea would be potentially effective trap crops for alectra as these have been confirmed to be non-host species (Phiri *et al.*, 2019). Sunflower has also proven effective as a rotation crop (Riches, 1989b). Care must be taken that the rotation crops do not facilitate other parasitic weeds, such as *Striga* spp. In Botswana (Riches, 1989b), Ethiopia (Hussien *et al.*, 2006) and Malawi (Kabambe *et al.*, 2008), *A. vogelii* co-occurs with *Striga asiatica*. If an alectra host, such as cowpea or groundnut, is rotated with a suitable *S. asiatica* host, such as sorghum, one parasitic weed problem will be replaced by another. In that respect, green manure species used for *Striga* spp. management, such as crotalaria (*Crotalaria grahamiana, C. juncea* and *C. ochroleuca*), could be used as these species have been identified as non-host species to alectra as well (Kabambe *et al.*, 2008). *Mucuna pruriens* was found to be immune to alectra by Kabambe *et al.* (2008)

in Malawi, but was previously observed by Riches (1989a) to host alectra in Botswana. This may be related to the differences in virulence between strains of alectra and therefore farmers need to be carefully advised on the use of trap crops.

Delayed crop establishment, effective against *Striga gesnerioides* in cowpea (Touré *et al.*, 1996) and against cereal-hosted *Striga* species (Gbèhounou *et al.*, 2004; Tippe *et al.*, 2017), could be considered. In Kenya, Bagnall-Oakeley *et al.* (1991) observed reduced alectra infection levels in cowpea following delayed sowing, but other studies indicate that results may be specific to the location/environment (Parker and Riches, 1993). This measure should therefore be tested locally, and through multiple cropping seasons, before recommendations can be made to farmers. The practice of late sowing has risks under rainfed crop production environments. An early-maturing crop variety would facilitate this practice as well as postharvest weed control methods; the less time a crop is in the field, the less it is exposed to stresses such as drought or alectra infection. Vice versa, the shorter the period that an alectra plant can benefit from a host plant, the lower the parasite's seed production.

Control methods based on external inputs could be evaluated as well, in particular those contributing to crop performance, such as organic or mineral fertilizers. In cowpea, Maganin *et al.* (1992) found fewer *A. vogelii* infections following application of moderate levels (60 kg N ha^{-1}) of nitrogen fertilizer, but with associated reductions in crop yield. Mugabe (1983) showed that the impact from *A. vogelii* parasitism on cowpea is mitigated by soil application of phosphorus. A pre-planting application of manure (at 16 t ha^{-1}) resulted in reduced *A. vogelii* infection levels, which confirmed farmers' observations in Kenya (Bagnall-Oakeley *et al.*, 1991). However, little research has been done on the effect of fertilizers on *Alectra* spp. and this merits further study.

8.4.5 Chemical control

Pre-emergence application of imazaquin (at 0.4 kg ha^{-1}) or pendimethalin (at 1 kg ha^{-1}) have proven the only effective *A. vogelii* herbicide solutions (Polniaszek and Parker, 1987).

8.4.6 Biological control

Like in witchweed, *Fusarium oxysporum* and *F. solani* prevent alectra seed production (Riches, 1989b), and these fungal pathogens could be used to develop effective biological control technologies.

8.4.7 Integrated management

A combination of the above measures in an integrated management approach is recommended for durable control. Such integrated alectra management strategies need to be based on sufficient understanding of the parasite and be locally adapted. They need to consider the specific alectra strain, the local environment (e.g. rainfall timing and distribution) and the available and farmer-preferred resources (e.g. organic or mineral fertilizers, crop and rotation-crop or cover-crop seeds, and labour). Alectra prevention or control measures should ideally also facilitate or include general weed management, as a range of weed species are known hosts to these parasites (see above).

References

Alonge, S.O., Lagoke, S.T.O. and Ajakaiye, C.O. (2001) Cowpea reactions to *Alectra vogelii* I: effect on growth. *Crop Protection* 20, 283–290.

Alonge, S.O., Lagoke, S.T.O. and Ajakaiye, C.O. (2005) Cowpea reactions to *Striga gesnerioides* II: effect on grain yield and nutrient composition. *Crop Protection* 24, 575–580.

Bagnall-Oakeley, H., Gibberd, V. and Nyongesa, T.E. (1991) The incidence and control of *A. vogelii* in Embu district, Kenya. In: Weber, H.C. and Forstreuter, W. (eds) *Proceedings of the 4th International Symposium on Parasitic Flowering Plants, Marburg, Germany, 1987*. Philipps-Universität, Marburg, Germany, pp. 53–66.

Beck, B.D.A. (1987) The effect of attack by *Alectra vogelii* Benth. on the yield of njugo beans (*Vigna subterranea* (L.) Verdc.). In: Weber, H.C. and Forstreuter, W. (eds) *Proceedings of the 4th International Symposium on Parasitic Flowering Plants, Marburg, Germany, 1987*. Philipps-Universität, Marburg, Germany, pp. 79–82.

Botha, P.J. (1946) Die lewensgeskiedenis morfologie en ontkiemingsfisiologie van *Alectra vogelii* Benth. PhD thesis, Potchefstroom University for Christian Higher Education, Potchefstroom, South Africa.

de la Harpe, A.C., Visser, J.H. and Grobbelaar, N. (1979) The chlorophyll concentration and photosynthetic activity of some parasitic flowering plants. *Zeitschrift für Pflanzenphysiologie* 95, 83–87.

Dieni, Z., Tignegre, J.B.D., Tongoona, P., Dzidzienyo, D., Asante, I.K. *et al.* (2018) Identification of sources of resistance to *Alectra vogelii* in cowpea *Vigna unguiculata* (L.) Walp. germplasm from Burkina Faso. *Euphytica* 214: 234.

Dörr, I., Visser, J.H. and Kollmann, D. (1979) On the parasitism of *A. vogelii* Benth. (Scrophulariaceae) III. The occurrence of phloem between host and parasite. *Zeitschrift für Pflanzenphysiologie* 94, 427–439.

Gbèhounou, G., Adango, E., Hinvi, J.C. and Nonfon, R. (2004) Sowing date or transplanting as components for integrated *Striga hermonthica* control in grain-cereal crops? *Crop Protection* 23, 379–386.

Gouws, J., Visser, J.H. and Grobbelaar, W.P. (1980) Some aspects of the bidirectional translocation of ^{14}C-labelled metabolites between *Alectra vogelii* Benth. and *Voandzeia subterranea* (L.) Thou. *Zeitschrift für Pflanzenphysiologie* 99, 223–233.

Horn, L., Shimelis, H. and Laing, M. (2015) Participatory appraisal of production constraints, preferred traits and farming system of cowpea in the northern Namibia: implications for breeding. *Legume Research* 38, 691–700.

Hussien, T., Mishra, B.B. and Gebrekidan, H. (2006) A new parasitic weed (*Alectra vogelii*) similar to *Striga* on groundnut in Ethiopia. *Tropical Science* 46, 139–140.

Kabambe, V., Katunga, L., Kapewa, T. and Ngwira, A.R. (2008) Screening legumes for integrated management of witchweeds (*Alectra vogelii* and *Striga asiatica*) in Malawi. *African Journal of Agricultural Research* 3, 708–715.

Kureh, I. and Alabi, S.O. (2003) The parasitic angiosperm *Alectra vogelii* (Benth.) can influence the growth and nodulation of host soybean (*Glycine max* (L.) Merrill). *Crop Protection* 22, 361–367.

Lagoke, S.T.O. (1989) Striga in Nigeria. In: Robson, T.O. and Broad, H.R. (eds) *Striga, Improved Management in Africa: Proceedings of the FAO/OAU All-Africa Government Consultation on Striga Control, Maroua, Cameroon, 20–24 October 1988*. FAO Plants Production and Protection Paper 96, IITA, Ibadan, pp. 68–73.

Maganin, I., Lagoke, S. and Adu, A. (1992) Nitrogen reduced Alectra emergence but depresses cowpea yield. *Striga Newsletter* 3, 13. FAO, Accra, Ghana.

Mainjeni, C.E. (1999) The host range of *Alectra vogelii* Benth. from Malawi and resistance in common bean and cowpea. MSc thesis, University of Bath, Bath, UK.

Mbwaga, A.M., Kaswende, J. and Shayo, E. (2000) *A Reference Manual on Striga Distribution and Control in Tanzania*. Kilosa Tanzania Ilonga Agricultural Research Institute, FARMESA, Harare, Zimbabwe.

Morawetz, J.J. and Wolfe, A.D. (2011) Taxonomic revision of the *Alectra sessiliflora* complex (Orobanchaceae). *Systematic Botany* 36, 141–152.

Mugabe, N.R. (1983) Effect of *Alectra vogelii* Benth. on cowpea (*Vigna unguiculata* (L.) Walp.) 1. Some aspects of reproduction of cowpea. *Zimbabwe Journal of Agricultural Research* 21, 135–147.

Müller, S., Hauck, C. and Schildknecht, H. (1992) Germination stimulants produced by *Vigna unguiculata* Walp cv Saunders Upright. *Journal of Plant Growth Regulation* 11, 77–84.

Musango, R., Pasipanodya, J.T., Tamado, T., Mabasa, S. and Makaza, W. (2022) *Alectra vogelii*: a threat to Bambara groundnut production under climate change: a review paper. *Journal of Agricultural Chemistry and Environment* 11, 83–105.

Musselman, L.J., Bolin, J.F. and Maass, E. (2020) Open sesame. *Haustorium* 78, 7–8.

Parker, C. (1988) Parasitic plants in Ethiopia. *Walia* 11, 21–27.

Parker, C. (2013) The parasitic weeds of the Orobanchaceae. In: Joel, D.M., Gressel, J. and Musselman, L.J. (eds) *Parasitic Orobanchaceae: Parasitic Mechanisms and Control Strategies*. Springer, Berlin, 313–344.

Parker, C. and Riches, C.R. (1993) *Parasitic Weeds of the World: Biology and Control*. CAB International, Wallingford, UK.

Phiri, C.K., Bokosi, V.H.K.J. and Mumba, P. (2019) Screening of *Alectra vogelii* ecotypes on legume and non-legume crop species in Malawi. *South African Journal of Plant and Soil* 36, 137–142.

Polniaszek, T.I. and Parker, C. (1987) Variation in host specificity of *Alectra vogelii* Benth. (Scrophulariaceae). In: Weber, H.C. and Forstreuter, W. (eds) *Proceedings of the 4th International Symposium on Parasitic Flowering Plants, Marburg, Germany, 1987*. Philipps-Universität, Marburg, Germany, pp. 613–619.

Polniaszek, T.I., Parker, C. and Riches, C.R. (1991) Variation in the virulence of *Alectra vogelii* populations on cowpea. *Tropical Pest Management* 37, 152–154.

Rambakudzibga, A.M., Manschadi, A.M. and Sauerborn, J. (2002) Host–parasite relations between cowpea and *Alectra vogelii*. *Weed Research* 42, 249–256.

Riches, C.R. (1989a) The biology and control of *A. vogelii* Benth. (Scrophulariaceae) in Botswana. PhD thesis, University of Reading, Reading, UK.

Riches, C.R. (1989b) The biology and control of witchweeds of grain legume and cereal crops in Botswana. In: *Research Projects 1983–1986 – Summaries of the Final Reports. First Programme Science and Technology for Development; Sub-programme: Tropics and Sub-Tropical Agriculture*. Commission of the European Communities, Brussels, pp. 318–322.

Riches, C.R., Hamilton, K.A. and Parker C. (1992) Parasitism of grain legumes by *Alectra* species. *Annals of Applied Biology* 121, 361–370.

Salako, E.A. (1984) Observations on the effect of *A. vogelii* infestation on the yield of groundnut. *Tropical Pest Management* 30, 209–221.

Samson, A. and Kehinde, A. (2009) Effect of *Alectra vogelii* and *Striga gesnerioides* infestations on the grain mineral elements' concentration of cowpea varieties. *Journal of Plant Protection Research* 49, 105–111.

Sauerborn, J., Müller-Stöver, D. and Hershenhorn, J. (2007) The role of biological control in managing parasitic weeds. *Crop Protection* 26, 246–254.

Singh, B.B. (2002) Breeding cowpea varieties for resistance to *Striga gesnerioides* and *Alectra vogelii*. In: Fatokun, C.A., Tarawali, S.A., Singh, B.B., Kormawa, P.M. and Tamò, M. (eds) *Challenges and Opportunities for Enhancing Sustainable Cowpea Production*. IITA Press, Ibadan, Nigeria, pp. 154–163.

Smith, D., Barkman, T.J. and dePamphilis, C.W. (2001) Hemiparasitism. In: Levin, S.A. (ed.) *Encyclopedia of Biodiversity, 2nd edn*. Academic Press, Orlando, Florida, pp. 70–78.

Tippe, D.E., Rodenburg, J., Van Ast, A., Anten, N.P.R., Dieng, I. *et al.* (2017) Delayed or early sowing: timing as parasitic weed control strategy in rice is species and ecosystem dependent. *Field Crops Research* 214, 14–24.

Touré, M., Olivier, A., Ntare, B.R. and St-Pierre, C.A. (1996) The influence of sowing date and irrigation of cowpea on *Striga gesnerioides* emergence. In: Moreno M.T., Cubero J.I., Berner, D., Joel, D., Musselman L.J. and Parker, C. (eds) *Advances in Parasitic Plant Research. Proceedings of the Sixth International Symposium on Parasitic Weeds*. Dirección General de Investigación Agraria, Córdoba, Spain, pp. 451–455.

Ugbaa, M.S., Osabuohien, O.L., Gowda B.S. and Timko, M.P. (2021) SSR markers associated with *Alectra vogelii* resistance gene in cowpea [*Vigna unguiculata* (L.) Walp.]. *Research Journal of Biotechnology* 16, 94–101.

Visser, J.H. (1978) The biology of *A. vogelii* Benth., an angiospermous root parasite. *Beiträge zur Chemischen Kommunikation in Bio- und* Ökosystemen, *Witzehausen* 1978, 279–294.

Visser, J.H. and Johnson, A.W. (1982) The effect of certain strigol analogues on the seed germination of Alectra. *South African Journal of Botany* 1, 75–76.

Visser, J.H., Dörr, I. and Kollmann, R. (1984) The "hyaline body" of the root parasite *Alectra orobanchoides* Benth. (Scrophulariaceae)–Its anatomy, ultrastructure and histochemistry. *Protoplasma* 121, 146–156.

Visser, J.H., Dörr, I. and Kollmann, R. (1990) Compatibility of *Alectra vogelii* with different leguminous host species. *Journal of Plant Physiology* 135, 737–745.

9 Broomrape

Abstract

Broomrapes are obligate holoparasitic weeds in a wide range of crop species, primarily vegetable crops, legumes, tobacco and sunflower. In Africa, broomrapes are mostly found in subtropical parts of the continent or at higher altitudes. Weedy broomrapes found in Africa are: *Orobanche crenata, O. cernua, O. minor, O. foetida, Phelipanche aegyptiaca* and *P. ramosa*. *Phelipanche aegyptiaca, Orobanche crenata, O. cernua* and *O. foetida* are restricted to Northern Africa, whereas *P. ramosa* and *O. minor* are also widely observed in Eastern and Southern Africa, with the exception of Madagascar. Most of these species, perhaps with the exception of *O. minor*, cause moderate to severe host damage and crop yield losses. Probably the most important, widespread and devastating broomrape species is *Phelipanche ramosa*. As obligate holoparasites, broomrapes are entirely dependent on their hosts for their germination but also for the acquisition of all their growth resources. Like witchweed and alectra, they have a rather long underground life-cycle stage during which they are not visible but significantly damage their host. They produce copious amounts of dust-like seeds. Broomrape can be managed by various means, including sanitation, solarization, the use of a limited range of herbicides, resistant crop varieties (although effective resistance is rare) and biological control agents. The most effective management of broomrape is likely to be achieved through crop rotations with trap or catch crops. As with other parasitic weeds, for long-term control, an integrated management approach is recommended.

9.1 Introduction

While less devastating and more restricted in Africa than witchweeds, broomrapes cause extensive damage to important food crops. The awkward English common name for these plants is derived from the Latin word for tuber or turnip because the non-green parasites were thought to be outgrowths of a common European shrub, *Cytisus*, known in English as broom. Broomrapes comprise two genera, *Orobanche* and *Phelipanche*, both known by the common name

 Parasitic Plants in African Agriculture (L.J. Musselman and J. Rodenburg)
DOI: 10.1079/9781789247657.0009

broomrape. Although the two genera can be separated on technical characters (not accepted by all plant taxonomists), noted below, they are very similar in biology, host impact and control.

They are serious parasites of vegetable crops such as tomato and carrot. The genus is centred in the Mediterranean region and is more temperate in its distribution than witchweeds. For this reason, broomrapes are not present at lower altitudes of the tropics. However, they can be devastating in crops grown at higher elevations or during the winter season in warmer climates.

The life history of broomrapes is complex and well documented in several sources. Extensive research has elucidated the elegant chemical communication between parasite and host, research that has resulted in the discovery of a new class of plant hormones known as strigolactones, reviewed in Xie *et al.* (2010). Studies of these compounds, their action on host and parasite and their possible role in control of *Striga* spp. and *Orobanche* spp. have elucidated the remarkable intricacy of these parasites, but to date have not led to effective control for the African smallholder.

From the farmer's standpoint, it is important to recognize broomrapes as more than weeds. They are parasites and for that reason need to be treated appropriately. By the time the broomrape emerges, irreparable host damage has been done. The seedlings are all subterranean so at the seedling stage the farmer is unaware that the crop is being parasitized.

9.2 Taxonomy and Identification

The taxonomy of the group is in a state of flux with several phylogenetic studies elucidating relationships. Botanists agree that it is a rapidly evolving group with species delineations less than those of *Striga* spp. Furthermore, the genus has historically suffered from extensive splitting into species and forms by earlier botanists. In addition, the inherent variability in the parasites has added to the confusion. Some of the variability is attributed to host influence, as some hosts support more rigorous growth than other hosts. An overview of the phylogeny is found in Schneeweiss *et al.* (2004), which is followed in this chapter.

9.3 Can a Native Broomrape Become a Problem?

There are numerous autochthonous holoparasitic members of the Orobanchaceae in Africa, especially in Southern Africa. This includes species of *Cistanche, Hyobanche* and others about which little is known of potential host ranges. But broomrapes previously considered benign can attack new hosts. In at least one case, the introduction of an alien weed, *Hypochaeris brasiliensis* (Asteraceae), resulted in the unprecedented spread of a native broomrape, *Orobanche uniflora*, in the south-eastern USA. Because the broomrape was attacking a weed, little attention was paid to the phenomenon (L.J. Musselman, personal observation). It is not impossible that this could occur where an introduced plant, a crop for example, could be attacked by a native parasite. This is what apparently happened when *Orobanche foetida* began to damage legume crops (see below).

9.4 Distribution

The following weedy broomrapes are known from Africa: *Orobanche crenata, O. cumana/O. cernua, O. minor, O. foetida, Phelipanche aegyptiaca* and *P. ramosa*. *Orobanche crenata, O. cernua* and *O. foetida* are restricted to Northern Africa, whereas *O. minor* is also widely observed in Eastern and Southern Africa, except for Madagascar (Fig. 9.1). The Egyptian broomrape, *Phelipanche aegyptiaca* is, true to its name, primarily observed in Egypt and other northern parts of the continent, whereas the distribution of *P. ramosa* extends to Sudan, Eritrea, Ethiopia, Tanzania and south to Namibia and South Africa (Fig. 9.2).

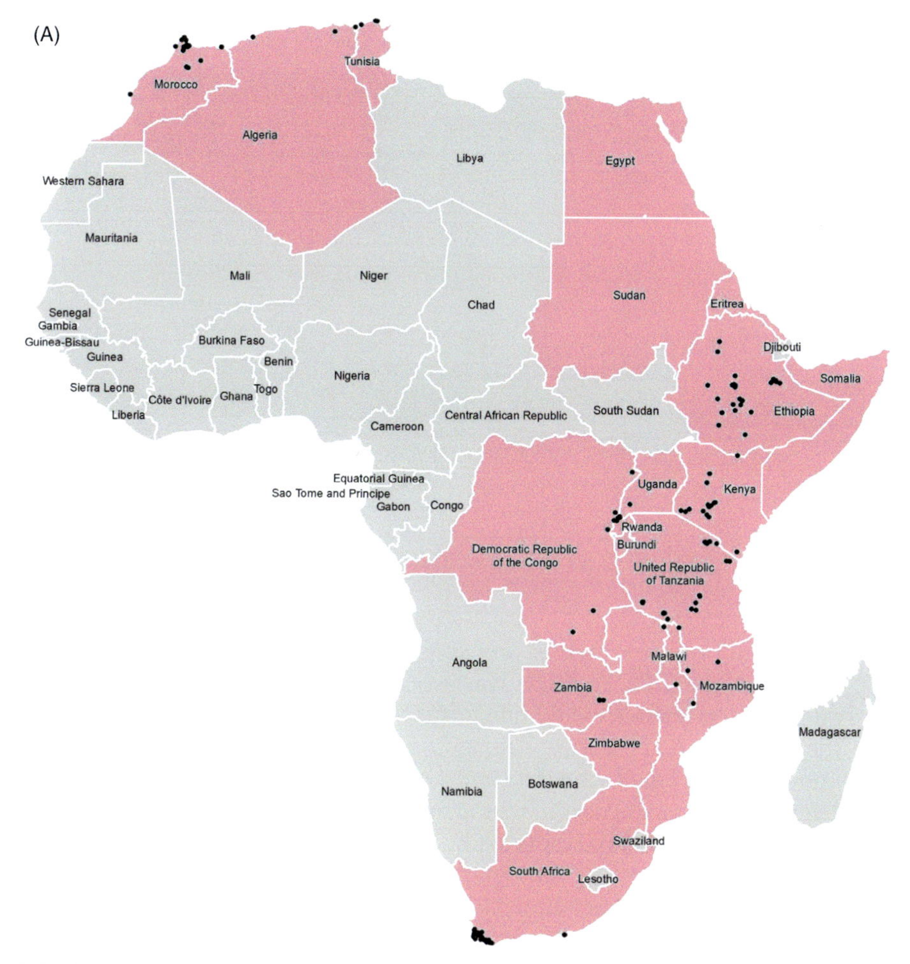

Fig. 9.1. Distribution in Africa of (A) *Orobanche minor*; (B) *O. crenata*, *O. foetida* and *O. cernua/O. cumana* (mapped distribution depends on extent of collections). Data from Global Biodiversity Information Facility: GBIF.org (accessed 9 September 2022), GBIF Occurrence Download (A) https://doi.org/10.15468/dl.qj8rmm; (B) https://doi.org/10.15468/dl.p267f9, https://doi.org/10.15468/dl.4z35r6 and https://doi.org/10.15468/dl.v4g8bh. Dots indicate georeferenced records from GBIF and non-grey colours indicate countries where the species has been observed.

Fig. 9.1. Continued.

Based on recent studies and consonant with the morphology of the parasites, the genus *Phelipanche* was erected to include those species with two additional structures called bracteoles that are at the base of the corolla. This means that two of the most damaging and widespread of the broomrapes once known as *Orobanche ramosa* and *O. aegyptiaca* are now known more accurately as *Phelipanche ramosa* (branched broomrape) (Fig. 9.3) and *P. aegyptiaca* (Egyptian broomrape), respectively. The difference between these two species is clearly explicated in Thorogood and Rumsey (2021). There are segregates of both species that are sometimes elevated to species, a fact not universally accepted by all botanists.

Orobanche crenata (Fig. 9.4) is a serious parasite with a broad host range in Africa. It is particularly damaging on faba bean in the Nile Valley of Egypt and Sudan, and also in Ethiopia (Gebru and Mesganaw, 2021). Minguez and Rubiales (2021) review the importance of this broomrape in faba beans and the need for crop improvement. Because it is commonly found on – but not restricted to – legumes, it is known as bean broomrape.

Fig. 9.2. Distribution of *Phelipanche aegyptiaca* and *P. ramosa* in Africa (mapped distribution depends on extent of collections). Data from GBIF.org (accessed 9 September 2022), GBIF Occurrence Download https://doi.org/10.15468/dl.k3454k and https://doi.org/10.15468/dl.v2t4a2. Dots indicate georeferenced records from GBIF and non-grey colours indicate countries where the species has been observed.

Orobanche cumana/*O. cernua* (Fig. 9.5) known in English as sunflower broomrape is widely distributed in the Mediterranean region, Eastern Europe, India, China and Northern Africa (Mokni and Domina, 2020). It is a constraint on the culture of sunflowers, tobacco and tomatoes in many places. Plants are usually unbranched and can be up to 1 m tall. The corollas are curved and usually blue at the orifice of the corolla. Flower colour can vary from yellow to purple. Seed production in broomrapes is enormous and sunflower broomrape may be the most fecund of them all. Most races are autogamous, that is, they pollinate themselves thus assuring abundant seed (L.J. Musselman, personal observation). A distinctive feature of this species is the usually single haustorium attaching the parasite to its hosts.

Fig. 9.3. *Phelipanche ramosa* showing the narrow bracteoles.

Parker (2009) notes reports of severe damage to tomatoes in Tanzania as well as a report from Niger. Later, Amri *et al.* (2012) noted infestation on sunflower (*Helianthus annuus*) in Tunisia. In pot studies, they found that the sunflower strain did not attack well-documented hosts including canola (*Brassica rapa*), lentil (*Lens culinaris*), faba bean (*Vicia fava*), pea (*Pisum sativum*), tomato (*Solanum lycopersicon*), capsicum pepper (*Capsicum annuum*) and potato (*Solanum tuberosum*). As noted earlier, the behaviour of broomrapes in pot studies differs from host parasitism in the field. This is particularly germane in the sunflower broomrape because of the well-studied evolution of races. This is summarized in Molinero-Ruiz *et al.* (2015). Intriguingly, *O. cumana* has not spread to North and South America or Australia despite extensive sunflower culture in those places. Its parasitism of tomatoes in Tanzania is a clear warning of its potential damage in other parts of Africa.

The taxonomy of this broomrape has been controversial. It is apparently derived from a widespread native species, *O. cernua,* which is not a pathogen. The putative differences between the two taxa are elaborated beyond practical utility in Pujadas-Salva and Velasco (2000) using morphological characteristics, which in broomrapes can be exceptionally variable and influenced by the host as well as seed fatty acids. The phylogeny of broomrapes strongly suggests that *O. cernua* and *O. cumana* are conspecific (Schneeweiss *et al.*, 2004).

Known by the common name of 'stinking broomrape' because of the foetid aroma of its flowers, *O. foetida* (Fig. 9.6) is of special interest because it has

Fig. 9.4. (A) *Orobanche crenata* in full flower. The flowers are pleasantly fragrant. (B) Parasitized faba bean on left, unparasitized control on right. (C) Parasitizing chickpea (*Cicer arietinum*) in Ethiopia.

been an agricultural problem only for the past three decades (Kharrat *et al.*, 1992). It is widespread in southern Europe where it has never been a pest. However, it is an important constraint on the cultivation of faba bean, *Vicia fava*, in Tunisia with crop losses of up to 90% (Boukteb *et al.*, 2021). In 2007, it was reported for the first time as a parasite in a Moroccan crop, common vetch, *Vicia sativa* (Rubiales *et al.*, 2005).

Unlike other species of broomrapes, *O. foetida* has been documented to parasitize plants of only one family, the Fabaceae (=Leguminosae). This host specificity could be a preadaptation to becoming a pathogen as a segetal weed.

Fig. 9.5. (A) Sunflower broomrape (*Orobanche cumana/O. cernua*) infestation in sunflower in Bulgaria. (B) Close up of flowers. The white strip is 1.0 cm. This population had considerable purple corolla coloration. (C) Attacking tomato in Jordan. The withered leaves, resembling frost damage, are a characteristic host response to parasitism. (D) Reported for no other broomrape, *O. cumana* has a single, large haustorial connection to the host. (E) Some seeds from a single plant of *O. cernua*.

Fig. 9.6. *Orobanche foetida* in southern Portugal. Purple is the most frequently encountered colour form. Some populations have yellowish flowers. Image courtesy of Chris Thorogood.

Its spread to common vetch in Morocco is a clear indication of its potential as a problem to leguminous crops in the Maghreb and other parts of its range.

Based on the extensive work on varieties and subspecies of *O. minor*, the research of Thorogood and Rumsey (2021) should be consulted to determine if reports of parasitism by this broomrape are indeed that species. Like other broomrapes, the plant size is influenced by the host. Colour can likewise vary (Fig. 9.7).

The host range of small broomrape is not as broad as some of its congeners. Two important crops are favoured, i.e. tobacco (*Nicotiana tabacum*) and carrot (*Daucus carota*). However, the greatest economic impact is on forage clovers (*Medicago* spp. and *Trifolium* spp.). But it can grow on *Abelia* spp. (ornamental shrubs in the Caprifoliaceae) and lettuce (*Lactuca sativa,* Asteraceae) so has potential for attacking a diversity of hosts. In common with other broomrapes, small broomrape will grow on hosts in pots that are not hosts in nature.

Fig. 9.7. (A) Small broomrape (*Orobanche minor*) on *Xanthium strumarium*, a widespread weed, Tigray State, Ethiopia. (B) Detail of the flower. (C) Parasitizing groundnut (*Arachis hypogaea*) in Ethiopia. The crop was severely damaged. (D) A lighter-coloured form on *Trifolium repens* (white clover), South Carolina. The colour of the corolla can vary from light yellow to purple.

No reportable *O. minor* damage to crops has been reported although the parasite is widespread on the continent occurring in Northern, Eastern and Southern Africa (Fig. 9.1). It has been found growing on groundnut (*Arachis hypogaea*) in Ethiopia (L.J. Musselman, personal observation). The *Flora of Egypt* states, cryptically, that small broomrape parasitizes 'species of Leguminosae, Compositae, and other plants' without further explication (Boulos, 2002).

Compared with *Phelipanche* spp., small broomrape is a minor problem in African agriculture, but agriculturalists should be made aware of its potential for serious damage.

Phelipanche aegyptiaca (Fig. 9.8) is frequently misidentified as *P. ramosa*, branched broomrape, which is understandable because the two species are morphologically similar, have sympatric ranges and attack many of the same hosts. In terms of host damage there appears to be little difference, and the same is true of control measures. Because of the taxonomic confusion between these two species and their inherent variability, it is likely that misidentification has occurred. Table 9.1 provides diagnostic guidance to distinguish between the two. Bearing their resemblance in mind, records indicate that the Egyptian broomrape has not spread as much as its congener. It was reported in California in 2014 and although it has not spread, there remains a very real potential of it becoming a parasite in tomatoes and other crops (Kelch, 2015; PPQ, 2018). Control measures for *Phelipanche aegyptiaca* and *P. ramosa* in other crops are basically the same (Goldwasser *et al.*, 2001).

Known in English by the common names of branched broomrape or hemp broomrape, *P. ramosa* (Fig. 9.9) is arguably the most devastating and certainly the most widespread of any broomrape occurring from the Mediterranean region to North and South America and India as well as Africa. Unlike the very similar *P. aegyptiaca*, branched broomrape is a component of the native flora of western Asia. Several taxa cluster around *P. ramosa* (cf. Schneeweiss *et al.*, 2004) and reports of infestations by the segregate taxa *P. mutelii* and *P. nana* may, in fact, be *P. ramosa*.

Phelipanche ramosa has the broadest host range of any broomrape, attacking crop species such as cauliflower, cabbage, celery, faba bean, lentils, hemp, tobacco, tomato, potato, aubergine, melon, mustards (including winter oilseed rape) and floricultural species, as well as many weedy and native species. The number of suitable hosts will increase with time. Stojanova *et al.* (2018) have found strong evidence of race formation in *P. ramosa* independent of geography or hosts. This invites comparison with the well-studied races of *O. cernua* on sunflower. Qasem (2022) provides extensive documentation

Table 9.1. Diagnostic characteristic summary for the two *Phelipanche* species.

Characteristic	*P. aegyptiaca*	*P. ramosa*
Corolla length	3–5 cm	2–4 cm
Anthers	Densely villous, often matted	Glabrous or few hairs
Floral fragrance	Fragrant	No obvious odour
Constriction near top of calyx	Absent	Present

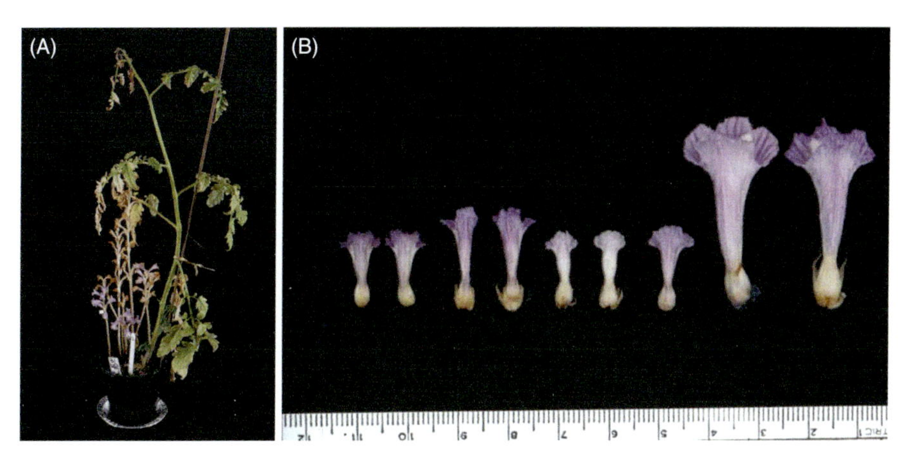

Fig. 9.8. (A) Egyptian broomrape (*Phelipanche aegyptiaca*) parasitizing tomato. (B) Seven corollas of *P. ramosa*, and two (extreme right) of *P. aegyptiaca.*

Fig. 9.9. (A) Branched broomrape (*Phelipanche ramosa*) on *Malva neglecta* (Malvaceae), Palestinian Territories. The much-branched characteristic is evident. (B) Closeup of flowers showing the constriction in the corolla above the calyx, Virginia, USA. (C) Tobacco infected with *P. ramosa*, Bulgaria. (D) Tomato crop devastated by branched broomrape, Sudan. (E) Seedlings ('tubercles'). Youngest are above. Broomrapes have no typical roots; the roots seen are from the host.

of hosts in Jordan. The diversity of hosts in this small country is indicative of the parasite's host latitude.

Phelipanche ramosa is apparently not native to Africa but has been widely spread in the Nile Valley and Ethiopia where in both regions it is a serious or even devastating parasite of tomato. In Morocco it has recently been found parasitizing turnip (Chedadi *et al.*, 2021).

According to Musselman (1984) branched broomrape could have been present in Sudan since 1904. More recently, it was reported at the Hudeiba Agricultural Research State in Northern Nile Province and seven years later had reached 'epidemic proportions' in that province.

9.5 Control

9.5.1 Sanitation

Like most weeds, broomrapes are spread by human activity. A small infestation of a hundred plants can yield millions of seeds, propagules that can remain viable in the soil for many years. There have been several cases where growing a crop had to stop because of the parasite.

An example of the introduction and spread of broomrape is reviewed in Belay *et al.* (2020) who report that *O. crenata* was first noted from Ethiopia in 1983 and is postulated to have been spread via contaminated seed exchange, grazing and movement on farm equipment.

For these reasons attention must be paid to the danger of transporting parasite seed and every effort possible should be made to remove the broomrape before the seeds mature. While this seems logical, years are required to reduce the seed in the soil. And to be effective every seed-producing plant must be removed because a single broomrape can produce hundreds of thousands of seeds.

9.5.2 Chemical control

There have been extensive studies on the control of broomrapes, especially sunflower broomrape, bean broomrape and Egyptian broomrape. Several herbicides have been successful in controlling *O. cumana* including imazapic and imidazolinone (Parker, 2009).

Information on the complex relationship of the host–parasite union and the number of effective herbicides is limited. Sulfonylurea herbicides can be effective against imbibed and germinated seeds. Herbicide control for broomrapes is summarized in Chapter 13 of this volume.

9.5.3 Cultural control

Witchweeds and broomrapes both require a root exudate to germinate. Many plants that do not serve as hosts can stimulate germination without supporting parasitism, these are known as trap crops. There is a diversity of species that can serve as trap crops. An exhaustive summary of trap crops is presented in Cartry *et al.* (2021). Although less frequently used, catch crops can also be used to reduce the number of seeds in the seed bank. Catch crop species are host plants that are not crops. They can be sown to germinate parasites and allow parasitism before they are destroyed and ploughed in. This needs to be done well before the parasites flower.

One form of trap/catch management of parasites is shifting cultivation in which a field is abandoned for several years before being planted with a crop. This was practised in Sudan for many years but with increased population and the need for crops, it is no longer practical (L.J. Musselman, personal observation). These approaches for control have little appeal to the smallholder constrained by available land and the need for subsistence crops. Trap crops and catch crops are more frequently used in the control of witchweed but there is increasing attention being paid to their use in broomrape control.

Solarization involves heating the soil surface with natural light producing temperatures under plastic sheets high enough to kill seeds. It has been used effectively against both *O. cernua/O. cumana* and *P. ramosa* in tomato in Ethiopia (Sahile *et al.*, 2005). Solarization can also be used in greenhouses (Mauro *et al.*, 2015).

9.5.4 Biological control

The search for effective biocontrol has extended for many years with only modest outcomes. To date the only work in Africa has been in Ethiopia with *O. cernua* and *Phelipanche ramosa* on tomato; in Egypt on faba beans and carrot parasitized by *O. crenata*; and in Morocco with *O. crenata* on several legumes. All these studies utilized the dipteran *Phytomyza orobanchia*. The larvae bore into the stem and developing capsules causing a reduction of up to 80% of seeds (Klein and Kroschel, 2002).

9.5.5 Genetic control

True resistance is extremely rare and may not exist. Developing a crop resistant to broomrapes is well suited to the smallholder if resistance is durable through several generations of seeds. This can be done by traditional means (Abdallah *et al.*, 2020) or through identifying genes and their activity (Boukteb *et al.*, 2021). With the increasing availability of CRISPR in a variety of applications, we can expect gene editing to develop resistant/tolerant host crops.

More detailed descriptions of broomrape management are provided in Chapter 13, this volume.

References

Abdallah, F., Kumar, S., Amri, A., Mentag, R., Kehel, Z. *et al.* (2020) Wild *Lathyrus* species as a great source of resistance for introgression into cultivated grass pea (*Lathyrus sativus* L.)

against broomrape weeds (*Orobanche crenata* Forsk. and *Orobanche foetida* Poir.). *Crop Science* 6, 263–276.

Amri, M., Abbes, Z., Youssef, S., Bouhadida, M., Salah, H. *et al.* (2012) Detection of the parasitic plant, *Orobanche cumana* on sunflower (*Helianthus annuus* L.) in Tunisia. *African Journal of Biotechnology* 11, 4163–4167.

Belay, G., Tesfaye, K., Hamwieh, A., Ahmed, S., Dejene, T. *et al.* (2020) Genetic diversity of *Orobanche crenata* populations in Ethiopia using microsatellite markers. *International Journal of Genomics* 2020: 3202037.

Boukteb, A., Sakaguchi, S., Ichihashi, Y., Kharrat, M., Nagon, A.T. *et al.* (2021) Analysis of genetic diversity and population structure of *Orobanche foetida* populations from Tunisia using RADseq. *Frontiers in Plant Science* 12: 618245.

Boulos, L. (2002) *Flora of Egypt,* Vol. 3, *(Verbenaceae-Compositae)*. Al Hadara Publishing, Cairo.

Cartry, D., Steinberg, C. and Gilbot-Leclerc, S. (2021) Main drivers of broomrape regulation. A review. *Agronomy for Sustainable Development* 41: 17.

Chedadi, T., Idrissi, O., Elkhabli, A., Khachtib, Y., Haddioui, A. *et al.* (2021) First report of branched broomrape (*Orobanche ramosa*) on turnip (*Brassica rapa*) in Morocco. *Plant Health Progress* 22, 92–93.

Gebru, A. and Mesganaw, M. (2021) Improving faba bean production and productivity through the integrated management of *Orobanche crenatae* at Kutaber, Amhara Region, Ethiopia. *International Journal of Agricultural Science and Food Technology* 7, 114–117.

Goldwasser, Y., Eizenberg, H., Hersehnhorn, J., Pakhine, D., Blumenfeld, T. *et al.* (2001) Control of *Orobanche aegyptiaca* and *O. ramosa* in potato. *Crop Protection* 20, 401–410.

Kelch, D. (2015) *Egyptian Broomrape/Orobanche aegyptiaca Pers.* California Department of Food and Agriculture, Sacramento, California.

Kharrat, M., Halifa, M.M., Linke, K.H. and Hadda, T. (1992) First report of *Orobanche foetida* Poiret on faba bean in Tunisia. *FABIS Newsletter* 30, 46–47.

Klein, O. and Kroschel, J. (2002) Biological control of *Orobanche* spp. with *Phytomyza orobanchia*, a review. *BioControl* 47, 245–277.

Mauro, R.P., LoMonaco, A., Lombardo, S., Restuccia, A. and Mauromicale, G. (2015) Eradication of *Orobanche/Phelipanche* spp. seedbank by soil solarization and organic supplementation. *Scientia Horticulturae* 193, 62–68.

Minguez, M.I. and Rubiales, D. (2021) Faba bean. In: Sadras, S.O. and Calderini, D.F. (eds) *Crop Physiology Case Histories for Major Crops*. Academic Press, London, pp. 452–481.

Mokni, R. and Domina, G. (2020) Additions to terrestrial flora of Tunisia: occurrence and taxonomic note. *Check List* 16, 553–561.

Molinero-Ruiz, L., Delavault, P., Perez-Vich, B., Pacureanu-Joita, M., Bulos, M. *et al.* (2015) History of the race structure of *Orobanche cumana* and the breeding of sunflower for resistance to this parasitic weed: a review. *Spanish Journal of Agricultural Research* 13: e10R01.

Musselman, L.J. (1984) Some parasitic angiosperms of Sudan: Hydnoraceae, Orobanchaceae, and *Cuscuta* (Convolvulaceae). *Notes from the Royal Botanic Garden, Edinburgh* 42, 21–38.

Parker, C. (2009) Observations on the current status of *Orobanche* and *Striga* problems worldwide. *Pest Management Science* 65, 453–459.

PPQ (2018) *Weed risk assessment for* Phelipanche aegyptiaca *(Pers.) Pomel (Orobanchaceae) – Egyptian broomrape*. United States Department of Agriculture, Animal and Plant Health Inspection Service, Plant Protection and Quarantine (PPQ), Raleigh, North Carolina. Available at: www.aphis.usda.gov/plant_health/plant_pest_info/weeds/downloads/wra/phelipanche-aegyptiaca.pdf (accessed 16 July 2023).

Pujadas-Salva, A.J. and Velasco, L. (2000) Comparative studies on *Orobanche cernua* L. and *O. cumana* Wallr. (Orobanchaceae) in the Iberian Peninsula. *Botanical Journal of the Linnean Society* 134, 513–527.

Qasem, J.R.S. (2022) *Parasitic Weeds of Jordan: Species, Hosts, Distribution and Management. Part 1. Root Parasites: Orobanchaceae, Santalaceae & Cynomoriaceae.* Bentham, Sharjah, United Arab Emirates.

Rubiales, D., Sadiki, M. and Román, B. (2005) First report of *Orobanche foetida* on common vetch (*Vicia sativa*) in Morocco. *Plant Disease* 89, 528.

Sahile, S., Abebe, G. and Al-Tawaha, A.M. (2005) Effect of soil solarization on *Orobanche* soil seed bank and tomato yield in Central Rift Valley of Ethiopia. *World Journal of Agricultural Sciences* 1, 143–147.

Schneeweiss, G.M., Colwell, A., Park, J., Jang, C. and Stuessy, T.F. (2004) Phylogeny of holoparasitic *Orobanche* (Orobanchaceae) inferred from nuclear ITS sequences. *Molecular Phylogenetics and Evolution* 30, 465–478.

Stojanova, B., Delourme, R., Duffé, P., Delavault, P. and Simier, P. (2018) Genetic differentiation and host preference reveal non-exclusive host races in the generalist parasitic weed *Phelipanche ramosa*. *Weed Research* 59, 107–118.

Thorogood, C. and Rumsey, F. (2021) *Broomrapes of Britain & Ireland. Botanical Society of Britain and Ireland Handbook No. 22.* Botanical Society of Britain and Ireland, Durham, UK.

Xie, X., Yoneyama, K. and Yoneyama, K. (2010) The Strigolactone Story. *Annual Review of Phytopathology* 48, 93–117.

10 Thonningia

Abstract

Thonningia is the only weedy obligate root holoparasite that is not a broomrape. The genus *Thonningia* has just one species: *T. sanguinea*. The only aboveground part of *T. sanguinea* plants are their red-coloured flowers. Thonningia is a parasitic weed on woody hosts such as rubber, coffee, cacao, oil palm and cassava, but the extent of damage it causes to its host is not well documented. It is mainly found in the forest and Guinea savannah zones of Western Africa, with eastern extensions into Ethiopia and southern extensions into Angola and Zambia, but because of its inconspicuous nature it may be easily overlooked and therefore data on the distribution of the species are presumably incomplete. There is no knowledge on effective thonningia management and the only way to control it is by excavation and removal from the host roots.

10.1 Introduction

The unusual, strikingly beautiful holoparasite thonningia is endemic to Western and Central Africa where it is a component of rainforest vegetation. There is no widely used common name in English, so the plant is often known by its genus name, thonningia. Unlike English, African languages have many names for thonningia, in large part because of the widespread use of the plant for herbal remedies and food (Neuwinger, 1996; Pompermaier *et al.*, 2019; Gonzalez and Sato, 2022). It is known to parasitize the roots of several commercially important trees including rubber.

10.2 Taxonomy and Identification

This species can be confused with no other plant. Its appearance is stunning when the flowers break through the leaf litter (Fig. 10.1). Otherwise, the plant

 Parasitic Plants in African Agriculture (L.J. Musselman and J. Rodenburg)
DOI: 10.1079/9781789247657.0010

Fig. 10.1. (A) Male inflorescence of *Thonningia sanguinea*, in southern Nigeria. Note the size of the inflorescence in relation to the leaf litter. (B) Parasite attachment to roots of rubber tree, southern Nigeria. Inflorescences are not open. Rubber roots are dark brown and those of thonningia are lighter. Images used by permission of E.I. Aigbokhan.

behaves like a secret agent as it creeps below the soil and preys upon its neighbours. Thonningia is a member of the Balanophoraceae and is a monospecific genus with a single species, *Thonningia sanguinea*. It shares many characteristics of other species of Balanophoraceae (see Kuijt, 1969; Heide-Jørgensen, 2008). Several scientific names are extant, an indication of variability within the taxon (Imarhiagbe and Aigbokhan, 2019b).

This is a most remarkable plant because it is never evident until flowering. The species is dioecious, that is, there are male and female plants. The flowers are inconspicuous, tiny and aggregated into dense inflorescences surrounded by stiff, brightly coloured red bracts. These are up to 5 cm in diameter. Floral biology shows thonningia is pollinated by flies (Goto *et al.*, 2012). The ill-defined fruit is fleshy with tiny seeds. Diaspore of the fruit and seed is unknown.

Population genetics indicate the possible presence of cryptic species, that is, species that are morphologically indistinguishable yet genetically different (Imarhiagbe and Aigbokhan, 2019b). The greater mass of the parasite is underground, a series of fleshy structures that form haustoria on the roots of woody hosts. Several terms have been used to describe the underground parts and it is not clear if they are morphologically tubers, rhizomes or roots. This parasite forms a tissue comprising both host and parasite components, a chimera apparently unique in angiosperms (Idu *et al.*, 2002). Since the plant lacks any chlorophyll, all nutrition must come from the host.

10.3 African Distribution

Thonningia is widespread, especially in Western Africa, but occurs erratically across the continent to Ethiopia and south as far as Zimbabwe and Madagascar (Fig. 10.2). Because of its furtive behaviour, it is likely that many populations have never been noted, suggesting it is more frequent and widespread than reported.

Fig. 10.2. Distribution of *Thonninga sanguinea* in Africa (mapped distribution depends on extent of collections). Data from Global Biodiversity Information Facility: GBIF.org (accessed 25 September 2022), GBIF Occurrence Download https://doi.org/10.15468/dl.jc6fmu. Dots indicate georeferenced records from GBIF and non-grey colours indicate countries where the species has been observed.

10.4 Hosts of Agronomic Interest

Thonningia has a limited host range (Olanya and Eilu, 2009; Imarhiagbe and Aigbokhan, 2019a) and apparently attacks only woody hosts. Thonningia attacks a few woody crops, and the restricted host range is perhaps more a reflection of its ecological needs than its host selection. It is important to note that non-indigenous hosts are attacked and seriously damaged. Perhaps the most adversely affected crop is rubber (*Hevea brasiliensis*) in Western Africa, especially Ghana, Nigeria and Sierra Leone (Fig. 10.1B; Idu *et al.*, 2002; Imarhiagbe and Aigbokhan, 2019b). In the past several years, there have been increased plantings of rubber in Western Africa so further reports of parasitism can be expected. Other crops parasitized by thonningia include coffee (*Coffea arabica*), cassava (*Manihot esculenta*), cacao (*Theobroma cacao*) and oil palm (*Elaeis guineensis*). Little data are available on crop damage.

10.5 Control

The only means of control is to excavate the parasite and remove it from the host roots, preferably before it has produced any seeds.

References

Gonzalez, A.M. and Sato, H. (2022) *Parasitic Plants*. IntechOpen, Vienna.

Goto, R., Yamakoshi, G. and Matsuzawa, T. (2012) A novel brood-site pollination mutualism?: the root holoparasite *Thonningia sanguinea* (Balanophoraceae) and an inflorescence-feeding fly in the tropical rainforests of West Africa. *Plant Species Biology* 27, 164–169.

Heide-Jørgensen, H.S. (2008) *Parasitic Flowering Plants*. Brill, Leiden, The Netherlands.

Idu, M., Begho, E.R. and Akpaja, E.O. (2002) Anatomy of attachment of the root parasite *Thonningia sanguinea* Vahl on *Hevea brasiliensis*. *Indian Journal of Natural Rubber Research* 15, 33–35.

Imarhiagbe, O. and Aigbokhan, E.I. (2019a) Studies on *Thonningia sanguinea* Vahl (Balanophoraceae) in southern Nigeria. Range and host preference. *International Journal of Conservation Science* 10721–10732.

Imarhiagbe, O. and Aigbokhan, E.I. (2019b) Studies on *Thonningia sanguinea* Vahl (Balanophoraceae) in southern Nigeria. Patterns of genetic diversity and population structure within and between populations. *Makara Journal of Science* 23, 193–203.

Kuijt, J. (1969) *Parasitic Flowering Plants*. University of California Press, Berkeley, California.

Neuwinger, H.D. (1996) Balanophoraceae: *Thonningia sanguinea*. In: *African Ethnobotany: Poisons and Drugs*. CRC Press, Boca Raton, Florida, pp. 249–251.

Olanya, C.A. and Eilu, G. (2009) Host–parasite relations of an angiospermous root parasite (*Thonningia sanguinea* Vahl) in logged and unlogged sites of Budongo forest reserve, western Uganda. *African Journal of Ecology* 47, 328–334.

Pompermaier, L., Schwaige, S., Mawunu, M., Lautenschlaeger, T. and Stuppner, H. (2019) Development and validation of a UHPLC-DAD method for the quantitative analysis of major dihydrochalcone glucosides from *Thonningia sanguinea* Vahl. *Planta Medica* 85, 911–916.

11 Other Root Parasites

Abstract

One of the objectives of this book is to raise awareness of the potential for new parasitic plant problems. This could arise either through the introduction of parasites from native or exotic sources, or by crops being grown in new areas where parasites are native. Further consideration must be given to climate change, for example growing crops at higher elevations. Although the parasitic guild includes stem parasites, such as dodder, love vine and mistletoe, those with the most likely interaction with field crops are the root parasites. We provide three examples: *Sopubia parviflora*, *Micrargeria filiformis* and *Thesium* spp. For all these species, there is limited knowledge on their distribution, parasitism, biology, host range and effects, as well as control. *Sopubia parviflora* and *Micrargeria filiformis* of the Orobanchaceae family are presumably facultative hemiparasites in rice systems, with wetlands as their natural habitat. Their distribution extends from Western to Eastern and Southern Africa. *Micrargeria filiformis* seems to be the most widespread. Both species are presumably controlled in similar ways to rice vampire weed. *Thesium* spp. seem restricted to Northern Africa (*T. humile*) or Southern Africa (*T. resedoides*). Biological features and control options of *Thesium* spp. are largely unknown.

11.1 Orobanchaceae

A review of African root parasites using herbarium specimens and published floras records two members of the Orobanchaceae parasitizing rice in Western Africa. These are *Sopubia parviflora* and *Micrargeria filiformis* (there are no widely used common names for these in English). We found only one record in the scientific literature on *M. filiformis* as a weed, in Burkina Faso (Traoré and Maillet, 1992). *Micrargeria filiformis* is mainly found in wetlands and swamps. It is a slender and erect medium-sized (30–80 cm) plant with pale to darker pink, opposite flowers (Fig. 11.1). This species is widespread in sub-Saharan Africa (Fig. 11.2). Specimens in the East African Herbarium showed numerous

 Parasitic Plants in African Agriculture (L.J. Musselman and J. Rodenburg)
DOI: 10.1079/9781789247657.0011

Fig. 11.1. *Micrargeria filiformis* in Guinea. (A) Habit. (B) Detail of flower.

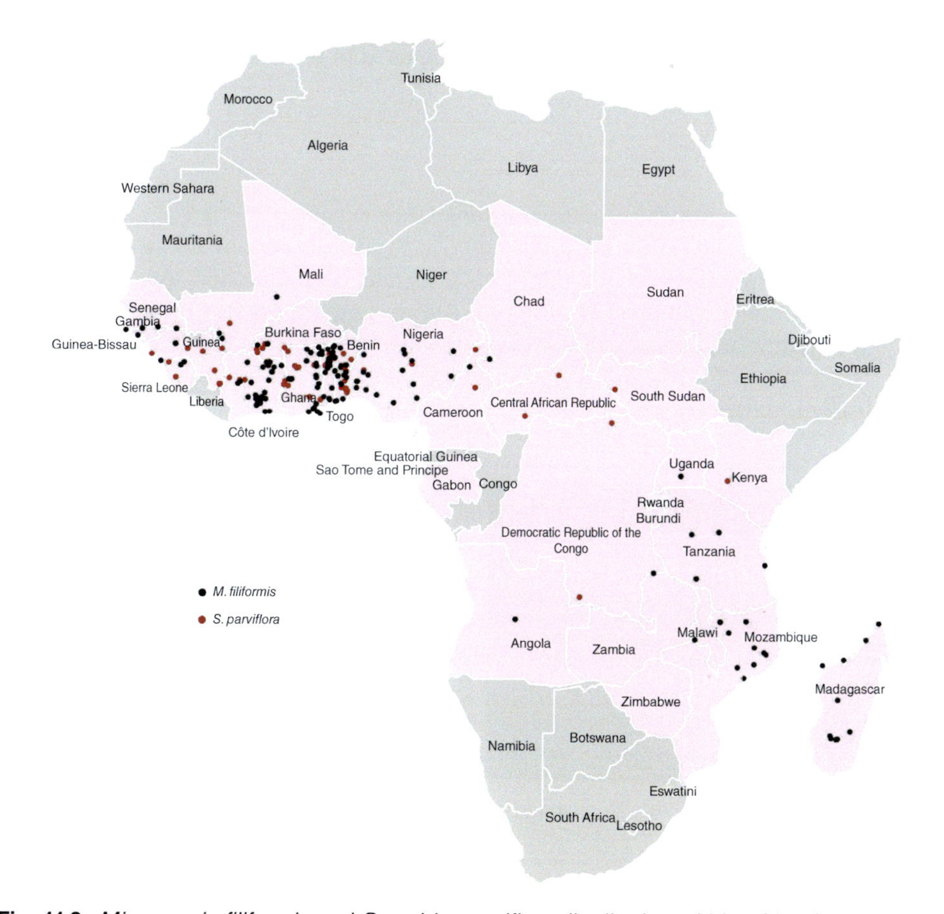

Fig. 11.2. *Micrargeria filiformis* and *Sopubia parviflora* distribution within Africa (mapped distribution depends on extent of collections). Data from Global Biodiversity Information Facility: GBIF.org (accessed 21 September 2022), GBIF Occurrence Download https://doi.org/10.15468/dl.4g4m3r and https://doi.org/10.15468/dl.4newu8. Blue and red dots indicate georeferenced records and non-grey colours indicate all countries where the species has been observed.

collections from the coastal regions of Kenya and Tanzania as well as lower elevations in the interior (L.J. Musselman, personal observation).

The distribution of *S. parviflora* is concentrated in Western Africa, with some observations from Angola, Zambia, Rwanda, Burundi, Kenya and Sudan (Fig. 11.2). It is found on wet, peaty soils (Gledhill, 1970) and is similar in appearance to *M. filiformis* but with off-white flowers (Fig. 11.3).

Micrargeria filiformis and *S. parviflora* both appear to be facultative hemiparasites, but this remains to be documented. These species could well become future weed problems in Africa. It is not unlikely that they have so far been overlooked because of their rather inconspicuous appearance and hence they could be more widely distributed and more frequently occurring in (rice) crops than reported here. There are no confirmed control measures against either of them, but it is expected that most measures that are effective against comparable hemiparasitic members of the Orobanchaceae, in particular rice vampire weed (*Rhamphicarpa fistulosa*), could be used to control both.

11.2 Thesiaceae

Lastly, species of the genus *Thesium* (Thesiaceae, sometimes placed in the Santalaceae), in particular *T. humile* observed in Northern Africa and *T. resedoides* in Southern Africa (Fig. 11.4), should be mentioned. At present they have not been reported as problems but as they are present on the continent, the weedy behaviour of these species in other parts of the world should be taken into consideration. Parasite biology and control of these species are not fully established.

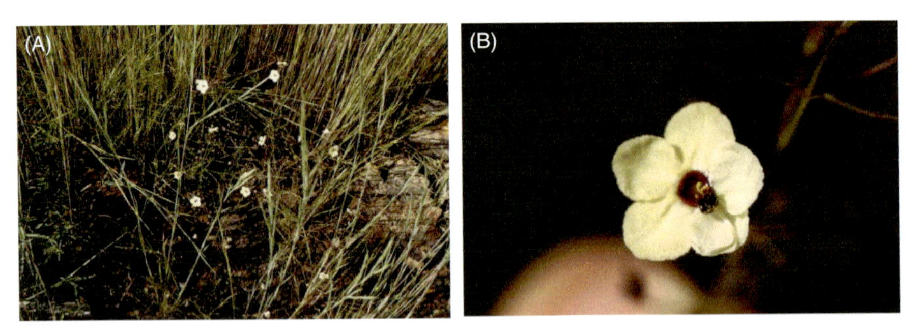

Fig. 11.3. (A) *Sopubia parviflora* at a margin of a rice field in Benin. (B) Detail of flower.

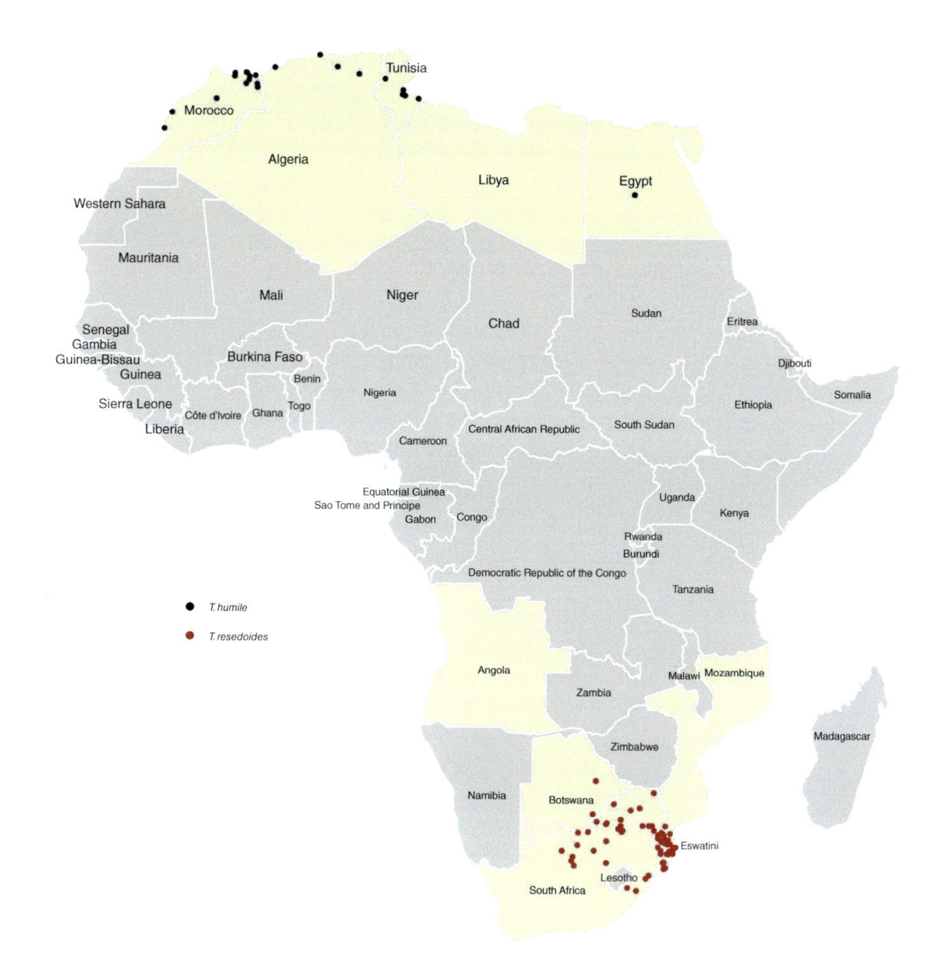

Fig. 11.4. *Thesium humile* and *T. resedoides* distribution within Africa (mapped distribution depends on extent of collections). Data from GBIF.org (accessed 25 September 2022), GBIF Occurrence Download https://doi.org/10.15468/dl.bvbect and https://doi.org/10.15468/dl.kkn9tt. Blue and red dots indicate georeferenced records and non-grey colours indicate all countries where the species has been observed.

References

Gledhill, D. (1970) Vegetation of superficial ironstone hardpans in Sierra Leone. *Journal of Ecology* 58, 265–274.

Traoré H. and Maillet J. (1992) Flore adventice des cultures céréalières annuelles du Burkina Faso. *Weed Research* 32, 279–293.

Part III Parasitic Plants in Practice

12 Parasitic Plants in African Smallholder Farming Systems

Abstract

Parasitic weeds in Africa are as diverse as the smallholder farming systems where they occur. Several factors may explain why parasitic weeds are such wicked problems in African agriculture, and why they require a broad range of effective control technologies. Parasitic weeds are well adapted to their hosts and some of their biological features make them particularly successful in crops. The environments where they cause problems are often already suboptimal for crop production, because of erratic rainfall and poor soil fertility. The affected farmers are often relatively resource-poor and have limited access to knowledge and means to manage parasitic weeds. In addition, there is often suboptimal communication between farmers – who are confronted with extant or newly emerging parasitic weed problems – and agricultural extension services, crop protection services and researchers – who could provide or investigate solutions. In those environments, farming systems, and agricultural extension and innovation systems, parasitic weeds are likely to pose a longer-term problem and impose a disproportional impact on crops and livelihoods.

12.1 Introduction

Parasitic weeds in African smallholder farming systems are clearly an important production constraint, particularly to the rainfed food crop production systems, including cowpea, maize, millet, rice and sorghum. Despite decades of research and development initiatives, including agricultural technology and knowledge transfer activities, parasitic weeds continue to impact crops and farmers' livelihoods in Africa (Ayongwa *et al.*, 2010; Scheiterle *et al.*, 2019) and there are no signs that the parasitic weed problem is decreasing. On the contrary, reports of newly emerging or increasing spread of parasitic weed species such as *Rhamphicarpa fistulosa* in rice (Rodenburg *et al.*, 2011b; Houngbedji *et al.*, 2014), *Striga hermonthica* in maize, sorghum, millet and rice systems in Western Africa (Dugje *et al.*, 2006; Aflakpui *et al.*, 2008) and *Alectra vogelii* in

 Parasitic Plants in African Agriculture (L.J. Musselman and J. Rodenburg)
DOI: 10.1079/9781789247657.0012

legume crops in Kenya (Karanja *et al.*, 2013), or alarming predictions of increasing invasions, such as *Cuscuta campestris, C. kilimanjari* and *C. reflexa* in mango, tea and coffee (Masanga *et al.*, 2021), indicate that the problem is worsening and becoming even more multi-faceted. For some of these parasitic weeds, an increasing range of control or management strategies is being developed (see Chapter 13, this volume) and yet smallholder farmers continue to have crop yield losses inflicted by parasitic weed. Figure 12.1 shows the multi-faceted character of this pest problem in African smallholder farming systems. In this chapter, we discuss three important reasons, partly linked or overlapping, for parasitic weeds being such difficult and persistent problems in Africa, needing a broad range of effective control technologies. These are: (i) biology of the parasites; (ii) farming systems and environments; and (iii) limited knowledge of or access to management options, partly due to suboptimal communication between researchers, crop protection services, agricultural extension services and smallholder farmers.

12.2 Biology of the Parasite

An important reason for the prevalence of root-parasitic weeds in African smallholder crops is biological. Root-parasitic weed seeds are very small (Fig. 12.2; Table 12.1) and most species have very high reproduction rates. In millet and

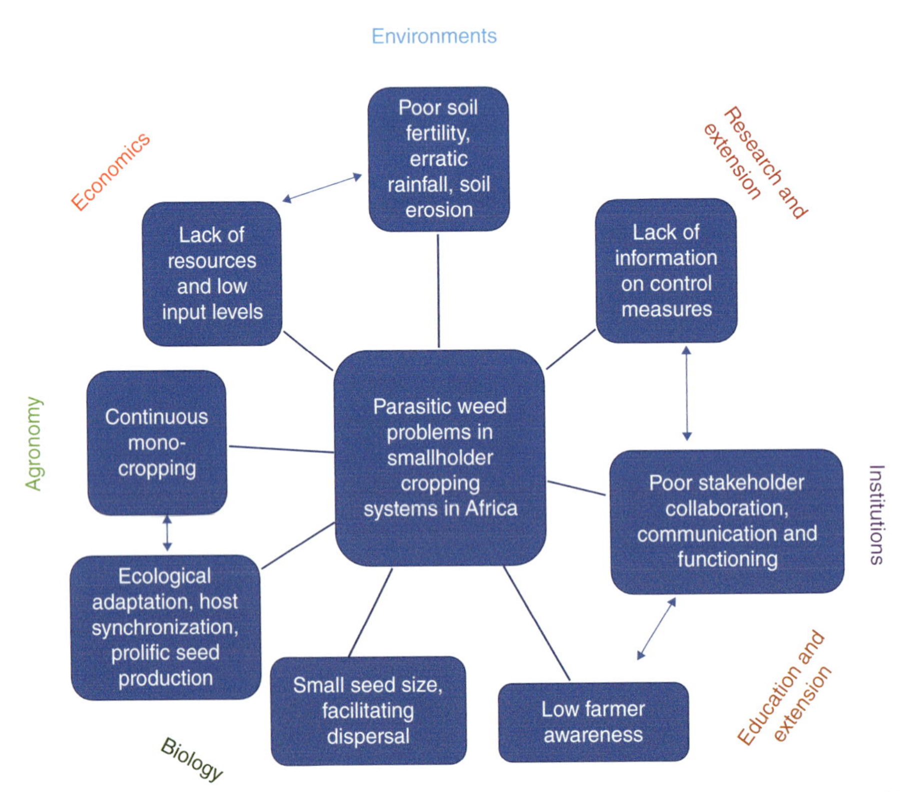

Fig. 12.1. The multi-faceted problem of parasitic weeds in smallholder farming systems in Africa.

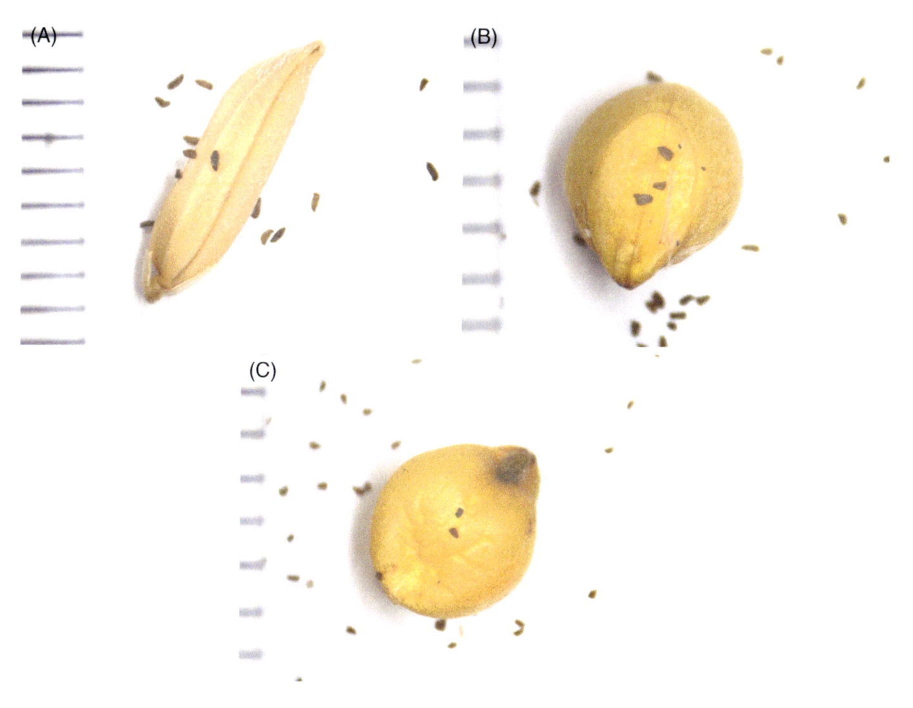

Fig. 12.2. Photos of parasitic weed seeds compared with seeds of their hosts (millimeter scale to the left): (A) *Rhamphicarpa fistulosa* seeds alongside a rice seed (variety IR64); (B, C) *Striga hermonthica* seeds (B) and *S. asiatica* seeds (C) alongside a sorghum seed (accession: IS10971).

Table 12.1. Seed sizes of some parasitic species of the Orobanchaceae.

Parasite	Seed length (mm)	Estimated seed weight (μg)
Obligate		
Striga hermonthica	0.38	4.5
S. asiatica	0.33	3.7
S. gesnerioides	0.33	5.8
S. aspera	0.30	5.5
Alectra vogelii[a]	0.20	0.3
Facultative		
Rhamphicarpa fistulosa	0.55	11.0
Buchnera hispida	0.55	–

[a]Excluding expanding seedcoat.
– : data unavailable.
Data from Parker and Riches, 1993; Rodenburg *et al.*, 2006, 2015a.

sorghum systems, for instance, seed production of *Striga hermonthica* has been estimated to range from 5000 (Webb and Smith, 1996) to 85,000 seeds per plant (Stewart, 1990). *Alectra vogelii* is even more prolific, with up to 600,000 seeds per plant (Botha, 1946; Visser, 1978). Farmers rarely remove or kill parasitic weeds after they have harvested the crop, allowing these parasites to continue producing flowers and seeds for as long as the remaining soil moisture permits. For *S. hermonthica*, for instance, postharvest seed production contributes

8%–39% of the total parasite reproduction capacity (Rodenburg *et al.*, 2006). This in turn allows a seasonally incrementing parasitic weed seed bank to build up in the soil, with associated increasing infection problems in subsequent crops.

Parasitic weeds often have seed biology and ecological features that make them particularly well adapted to crops, or difficult to manage. In tree crops, for instance, fruits of the stem-parasitic mistletoe may attract specific bird species that contribute to seed dispersal between trees (De Buen and Ornelas, 1999). For root-parasitic weeds, seed dormancy breaking and germination are often well synchronized with the presence of a suitable host (Fig. 12.3). In annual cropping systems, seeds of many parasitic weeds (e.g. witchweed, broomrape) require a preconditioning period and exposure to host root signals before they can germinate (Yoneyama *et al.*, 2010). Both mechanisms increase the likelihood of a nearby host plant and ensure that germination and further growth and development are well synchronized with that of the crop.

Annual root-parasitic plants also produce high numbers of very small seeds that are easily dispersed and often long-lived (Joel, 2013), contributing to a rapid build-up of a persistent seed bank in arable soils.

In addition to the prolific production of minute seeds, effective dissemination and the mechanisms ensuring germination only occur when the environmental conditions are right and a host plant is available, parasitic weed species (in particular, *S. hermonthica*) often have a high genetic variation (Welsh and Mohamed, 2011), and hence differences in virulence, within a population as well as between populations. This enables a population to quickly adapt to

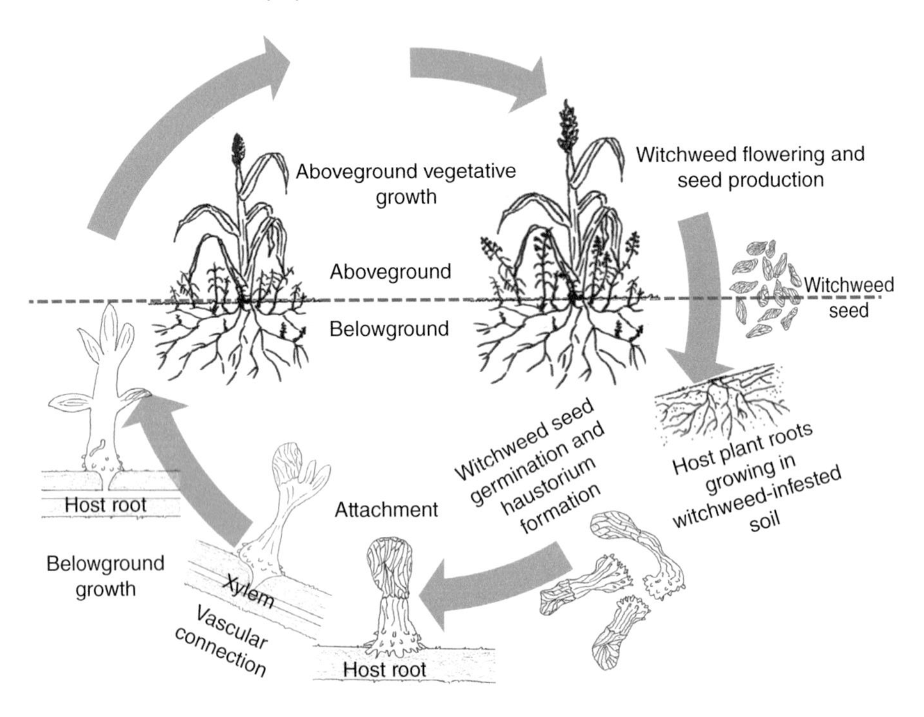

Fig. 12.3. Lifecycle of witchweeds, *Striga* spp. (Adapted from Mwangangi *et al.*, 2021.)

changing environmental, crop and management conditions. Crop genotypes that are resistant against a population in one location ('ecotype') may not be resistant against a population at another location (Rodenburg *et al.*, 2017). A wide range of virulence factors within a witchweed population have recently been identified, and this may enable individuals to overcome different host-plant resistance mechanisms present in a crop genotype (Qiu *et al.*, 2022), resulting in resistance breakdown over time.

12.3 Farming Systems and Environments

Parasitic weeds occur naturally in the non-agricultural vegetation and are therefore well adapted to the nearby crop environments where they can become weed problems. Witchweed problems seem to be most prevalent and most damaging on infertile soils (Ayongwa *et al.*, 2010; Kamara *et al.*, 2014) and in areas of erratic rainfall and intensified, mono-cropping, production systems (Cardwell and Lane, 1995; Dugje *et al.*, 2006; Dossou-Aminon *et al.*, 2016). The witchweed problem, both in terms of distribution and in terms of impact, is expected to increase due to the effects of climate change (Rodenburg *et al.*, 2011a; Dossou-Aminon *et al.*, 2016; Mudereri *et al.*, 2020; Kimathi *et al.*, 2022).

Parasitic weed problems are disproportionately affecting the poorest farmers (Kroschel, 1999; Parker, 2009). Not only do these farmers often crop the most marginal soils, but they also are more constrained when it comes to implementation of effective management technologies, because of the costs and availability of those (Tippe *et al.*, 2017). Moreover, they often have a range of other crops or economic endeavours that require their attention and investments (Giller *et al.*, 2021).

Farming systems across Africa are affected by parasitic weeds, but this is far from being a simple and well-defined problem. This is because a broad range of parasitic weed species often coexist on an individual farm. Moreover, there is a wide heterogeneity in biophysical and socio-economic environments among farms and rural areas in Africa (Vanlauwe *et al.*, 2019) and this further complicates the identification of suitable control strategies. In fact, what is required for better control measures of parasitic weeds in smallholder farming systems in Africa is to generate a 'basket of options' (analogous to Ronner *et al.*, 2021), allowing individual farmers to select the control technologies best fitting the specific biophysical and socio-economic conditions and production objectives of their farms. (For examples of effective management practices, see Chapter 13, this volume.) This would also facilitate the composition of locally adapted integrated parasitic weed management approaches, as no single technology can address this problem when used in isolation (Goldwasser and Rodenburg, 2013).

12.4 Knowledge and Access to Management Options

An important reason certain species of parasitic weeds are such persistent problems in smallholder cropping systems in Africa is that they are generally difficult to manage (see Chapter 13, this volume). This in turn can be explained from a biological or ecological as well as from a socio-economic and institutional perspective.

Parasitic plants are physically attached to the vascular system of the host plants by means of their haustoria (Yoshida *et al.*, 2016), inherently making it difficult to kill them without simultaneously harming their hosts. As noted earlier, the biology, ecology or seed dispersal of parasitic weeds is well adapted to that of their hosts (Těšitel, 2016). Such biological adaptations and interactions complicate the management of parasitic weeds (although some adaptations, such as the germination stimulant requirement, also offer control solutions – see Chapter 13, this volume). Some parasites, for example the mistletoes, make use of animal vectors (e.g. birds) for seed dispersal, and such vectors are nearly impossible to control. Other seed dispersal vectors, such as agricultural machines and equipment or free-browsing cattle, would be easier to control, but this would require awareness among farmers and protective measures such as fences or hedges around crop fields.

The effective control of parasitic weeds necessitates a long-term investment based on both technical and organizational approaches, as illustrated by the case of *Striga asiatica* in the USA. The discovery of *S. asiatica* in maize, sorghum and sugarcane production systems in North and South Carolina in 1956 led to the development of a well-coordinated and successful but very expensive programme of control and quarantine measures, dependent on functional crop protection and agricultural extension services (Eplee, 1992). Despite decades of control and quarantine restrictions and the expectation that *S. asiatica* would be completely eradicated by 1995 (Eplee, 1992), this has not yet been achieved, illustrating just how persistent parasitic weed problems are.

In smallholder cropping systems in Africa, the available resources often do not allow for a similar high-input approach. Suboptimally functioning markets and inaccessibility of information and extension and crop protection services in rural areas further hampers implementation and monitoring of control and prevention efforts. To serve African smallholder farmers, control measures need to be adapted to the available resources and the institutional and (input) market environments. In practice, this means that measures that heavily rely on industries and commercial products (e.g. agrochemicals, hybrid seeds, machines) or need specialist knowledge (e.g. herbicides, some biological control agents) may be less likely to be suitable.

Smallholder farmers in Africa often lack knowledge on control technologies or these technologies are not available, accessible or affordable to them, as shown by studies on witchweed management on smallholder farms (Mrema *et al.*, 2017; Tippe *et al.*, 2017). The lack of awareness and knowledge is mostly due to research–extension–farmer communication lines that have frequently been observed to be dysfunctional.

An important reason a parasitic weed problem such as *Rhamphicarpa fistulosa* was allowed to develop is that it was (and in many places still is) probably not recognized as a problem. In Benin, this parasite was observed in rice fields by researchers in the 1990s, but even years later extension agents were not aware of it (Rodenburg *et al.*, 2015b). Farmers were obviously aware but did not know how to control it. Similar observations on this species were made in Uganda (Rodenburg *et al.*, 2015a). Because of a lack of specific training, extension agents may also not be aware of the most effective and feasible control strategies to advise the farmers (Schut *et al.*, 2015a).

Parasitic weeds present complex agricultural problems that can only be solved through the concerted action of multiple stakeholders (Schut *et al.*, 2015b). The challenge is linked to poor socio-economic and institutional conditions; it is thus primarily a problem on farms of smallholders who are often poorly informed with limited financial means and production resources to control the weed. Part of this is caused by inadequate access to extension services, poor education, and increasing input prices or lack of functional markets in rural Africa (Schut *et al.*, 2015a).

12.5 Concluding Remarks Regarding Parasitic Weeds on Smallholder Farms

Parasitic weeds form an intricate and persistent problem due to their dispersal adaptations, germination biology, large numbers of minute seeds, high genetic heterogeneity within a population and the existence of many ecotypes with differing virulence.

Control technologies should be suitable and feasible for smallholder farmers who often have little financial scope and have many other concerns. Smallholder farmers do not, however, comprise a homogeneous and well-defined group; there is a huge variation between farmers within a specific country and between countries and subregions (Giller *et al.*, 2021). What works for one farmer does not necessarily work for another farmer. This may be due to differences in biophysical conditions (e.g. soil types, altitude, slopes, climate regimes), agronomy (e.g. crop species, cropping calendars, cropping systems) or farm technical or economic aspects (e.g. availability of livestock, mechanization and agro-inputs, availability of seeds, investment power, functioning markets).

The *best-bet* and *best-fit* approach to tackle parasitic weeds in smallholder farming systems in a durable way is by generating a 'basket of options' containing a wide range of control technologies and applying a locally adapted selection of these in an integrated manner, as it is well established that a single approach does not work.

Finally, the problem of parasitic weeds in African agriculture is intrinsically associated with poverty and related problems of availability of and access to technology and information. Addressing poverty and improving connection and communication between research and crop protection and agricultural extension services, as well as between agricultural extension services and farmers, will greatly contribute to a better control over parasitic weeds in smallholder crop production systems in Africa.

References

Aflakpui, G.K.S., Bolfrey-Arku, G.E.K., Anchirinah, V.M., Manu-Aduening, J.A. and Adu-Tutu, K.O. (2008) Incidence and severity of *Striga* spp. in the coastal savanna zone of Ghana. Results and implications of a formal survey. *Outlook on Agriculture* 37, 219–224.

Ayongwa, G.C., Stomph, T.J., Hoevers, R., Ngoumou, T.N. and Kuyper, T.W. (2010) Striga infestation in northern Cameroon: magnitude, dynamics and implications for management. *NJAS – Wageningen Journal of Life Sciences* 57, 159–165.

Botha, P.J. (1946) Die lewensgeskiedenis morfologie en ontkiemingsfisiologie van *Alectra vogelii* Benth. PhD thesis, Potchefstroom University for Christian Higher Education, Potchefstroom, South Africa.

Cardwell, K.F. and Lane, J.A. (1995) Effect of soils, cropping system and host phenotype on incidence and severity of *Striga gesnerioides* on cowpea in West Africa. *Agriculture, Ecosystems & Environment* 53, 253–262.

De Buen, L.L. and Ornelas, J.F. (1999) Frugivorous birds, host selection and the mistletoe *Psittacanthus schiedeanus*, in central Veracruz, Mexico. *Journal of Tropical Ecology* 15, 329–340.

Dossou-Aminon, I., Dansi, A., Ahissou, H., Cisse, N., Vodouhe, R. and Sanni, A. (2016) Climate variability and status of the production and diversity of sorghum (*Sorghum bicolor* (L.) Moench) in the arid zone of northwest Benin. *Genetic Resources and Crop Evolution* 63, 1181–1201.

Dugje, I.Y., Kamara, A.Y. and Omoigui, L.O. (2006) Infestation of crop fields by *Striga* species in the savanna zones of northeast Nigeria. *Agriculture, Ecosystems & Environment* 116, 251–254.

Eplee, R.E. (1992) Witchweed (*Striga asiatica*): an overview of management strategies in the USA. *Crop Protection* 11, 3–7.

Giller, K.E., Delaune, T., Silva, J.V., Descheemaeker, K., Van De Ven, G. *et al.* (2021) The future of farming: who will produce our food? *Food Security* 13, 1073–1099.

Goldwasser, Y. and Rodenburg, J. (2013) Integrated agronomic management of parasitic weed seed banks. In: Joel, D.M., Gressel, J. and Musselman, L.J. (eds) *Parasitic Orobanchaceae: Parasitic Mechanisms and Control Strategies.* Springer, Berlin, pp. 393–413.

Houngbedji, T., Pocanam, Y., Shykoff, J., Nicolardot, B. and Gibot-Leclerc, S. (2014) A new major parasitic plant in rice in Togo: *Rhamphicarpa fistulosa*. *Cahiers Agricultures* 23, 357–365.

Joel, D.M. (2013) Seed production and dispersal in the Orobanchaceae. In: Joel, D.M., Gressel, J. and Musselman, L.J. (eds) *Parasitic Orobanchaceae: Parasitic Mechanisms and Control Strategies.* Springer, Berlin, pp. 143–146.

Kamara, A.Y., Ekeleme, F., Jibrin, J.M., Tarawali, G. and Tofa, I. (2014) Assessment of level, extent and factors influencing *Striga* infestation of cereals and cowpea in a Sudan Savanna ecology of northern Nigeria. *Agriculture, Ecosystems & Environment* 188, 111–121.

Karanja, J., Nguluu, S.N., Wambua, J. and Gatheru, M. (2013) Response of cowpea genotypes to *Alectra vogelii* parasitism in Kenya. *African Journal of Biotechnology* 12, 6591–6598.

Kimathi, E., Abdel-Rahman, E.M., Lukhoba, C., Ndambi, A., Mudereri, B.T. *et al.* (2022) Ecological determinants and risk areas of *Striga hermonthica* infestation in western Kenya under changing climate. *Weed Research* 63, 45–56.

Kroschel, J. (1999) Analysis of the *Striga* problem, the first step towards further joint action. In: Kroschel, J., Mercer-Quarshie, H. and Sauerborn, J. (eds) *Advances in Parasitic Weed Control at On-farm Level*. Vol. I. *Joint Action to Control Striga in Africa*. Margraf Verlag, Weikersheim, Germany, pp. 3–25.

Masanga, J., Mwangi, B.N., Kibet, W., Sagero, P., Wamalwa, M. *et al.* (2021) Physiological and ecological warnings that dodders pose an exigent threat to farmlands in Eastern Africa. *Plant Physiology* 185, 1457–1467.

Mrema, E., Shimelis, H., Laing, M. and Bucheyeki, T. (2017) Farmers' perceptions of sorghum production constraints and *Striga* control practices in semi-arid areas of Tanzania. *International Journal of Pest Management* 63, 146–156.

Mudereri, B.T., Abdel-Rahman, E.M., Dube, T., Landmann, T., Khan, Z. *et al.* (2020) Multi-source spatial data-based invasion risk modeling of Striga (*Striga asiatica*) in Zimbabwe. *GIScience and Remote Sensing* 57, 553–571.

Mwangangi, I.M., Büchi, L., Haefele, S., Bastiaans, L., Runo, S. *et al.* (2021) Combining host plant defence with targeted nutrition: key to durable control of hemiparasitic *Striga* in cereals in sub-Saharan Africa? *New Phytologist* 230, 2164–2178.

Parker, C. (2009) Observations on the current status of *Orobanche* and *Striga* problems worldwide. *Pest Management Science* 65, 453–459.

Parker, C. and Riches, C.R. (1993) *Parasitic Weeds of the World: Biology and Control*. CAB International, Wallingford, UK.

Qiu, S., Bradley, J.M., Zhang, P., Chaudhuri, R., Blaxter, M. *et al.* (2022) Genome-enabled discovery of candidate virulence loci in *Striga hermonthica*, a devastating parasite of African cereal crops. *New Phytologist* 236, 622–638.

Rodenburg, J., Bastiaans, L., Kropff, M.J. and Van Ast, A. (2006) Effects of host plant genotype and seedbank density on *Striga* reproduction. *Weed Research* 46, 251–263.

Rodenburg, J., Meinke, H. and Johnson, D.E. (2011a) Challenges for weed management in African rice systems in a changing climate. *Journal of Agricultural Science* 149, 427–435.

Rodenburg, J., Zossou-Kouderin, N., Gbèhounou, G., Ahanchede, A., Touré, A. *et al.* (2011b) *Rhamphicarpa fistulosa*, a parasitic weed threatening rain-fed lowland rice production in sub-Saharan Africa – a case study from Benin. *Crop Protection* 30, 1306–1314.

Rodenburg, J., Morawetz, J.J. and Bastiaans, L. (2015a) *Rhamphicarpa fistulosa*, a widespread facultative hemi-parasitic weed, threatening rice production in Africa. *Weed Research* 55, 118–131.

Rodenburg, J., Schut, M., Demont, M., Klerkx, L., Gbehounou, G. *et al.* (2015b) Systems approaches to innovation in pest management: reflections and lessons learned from an integrated research program on parasitic weeds in rice. *International Journal of Pest Management* 61, 329–339.

Rodenburg, J., Cissoko, M., Kayongo, N., Dieng, I., Bisikwa, J. *et al.* (2017) Genetic variation and host-parasite specificity of *Striga* resistance and tolerance in rice: the need for predictive breeding. *New Phytologist* 214, 1267–1280.

Ronner, E., Sumberg, J., Glover, D., Descheemaeker, K.K.E., Almekinders, C.J.M. *et al.* (2021) Basket of options: unpacking the concept. *Outlook on Agriculture* 50, 116–124.

Scheiterle, L., Haring, V., Birner, R. and Bosch, C. (2019) Soil, *Striga*, or subsidies? Determinants of maize productivity in northern Ghana. *Agricultural Economics* 50, 479–494.

Schut, M., Rodenburg, J., Klerkx, L., Hinnou, L.C., Kayeke, J. *et al.* (2015a) Participatory appraisal of institutional and political constraints and opportunities for innovation to address parasitic weeds in rice. *Crop Protection* 74, 158–170.

Schut, M., Rodenburg, J., Klerkx, L., Kayeke, J., Van Ast, A. *et al.* (2015b) RAAIS: Rapid Appraisal of Agricultural Innovation Systems (Part II). Integrated analysis of parasitic weed problems in rice in Tanzania. *Agricultural Systems* 132, 12–24.

Stewart, G. (1990) Witchweed: a parasitic weed of grain crops. *Outlook on Agriculture* 19, 115–117.

Těšitel, J. (2016) Functional biology of parasitic plants: a review. *Plant Ecology and Evolution* 149, 5–20.

Tippe, D.E., Rodenburg, J., Schut, M., Van Ast, A., Kayeke, J. *et al.* (2017) Farmers' knowledge, use and preferences of parasitic weed management strategies in rain-fed rice production systems. *Crop Protection* 99, 93–107.

Vanlauwe, B., Coe, R.I.C. and Giller, K.E. (2019) Beyond averages: new approaches to understand heterogeneity and risk of technology success or failure in smallholder farming. *Experimental Agriculture* 55, 84–106.
Visser, J.H. (1978) The biology of *A. vogelii* Benth., an angiospermous root parasite. *Beiträge zur Chemischen Kommunikation in Bio- und* Ökosystemen, *Witzehausen,* 279–294.
Webb, M. and Smith, M.C. (1996) Biology of *Striga hermonthica* (Scrophulariaceae) in Sahelian Mali: effects on pearl millet yield and prospects of control. *Weed Research* 36, 203–211.
Welsh, A.B. and Mohamed, K.I. (2011) Genetic diversity of *Striga hermonthica* populations in Ethiopia: evaluating the role of geography and host specificity in shaping population structure. *International Journal of Plant Sciences* 172, 773–782.
Yoneyama, K., Awad, A.A., Xie, X.N. and Takeuchi, Y. (2010) Strigolactones as germination stimulants for root parasitic plants. *Plant and Cell Physiology* 51, 1095–1103.
Yoshida, S., Cui, S.K., Ichihashi, Y. and Shirasu, K. (2016) The haustorium, a specialized invasive organ in parasitic plants. *Annual Review of Plant Biology* 67, 643–667.

13 Parasitic Weed Management

Abstract

Parasitic weeds can be managed through sanitation, mechanical, chemical, cultural, biological or crop-varietal methods. Because of the high genetic heterogeneity, fecundity and excellent environmental and crop adaptations, none of these methods is likely to provide effective, long-term control as a stand-alone technology. Management of parasitic weeds is therefore most durably achieved by an integrated approach that combines multiple complementary technologies. The number of suitable technologies that could be integrated this way is more restricted for stem parasites than for root parasites. For stem parasites, the main intervention is prevention of parasitism, including early, pre-attachment chemical control. Once infection has taken place, the parasite can only be controlled by removal, or by pruning infected branches for mistletoe on tree crops. Root parasites can best be managed by crop rotations or intercropping, the use of resistant varieties and manual removal of parasitic plants before they flower.

A range of biological, cultural and chemical control technologies have been developed in recent years. For many minor or emerging parasitic weed species, the management options are limited. It is, however, expected that many principles and mechanisms underlying control technologies against major parasitic weed species would also apply to the minor ones. Apart from the differentiation between stem and root parasites, it is important for the selection of effective control methods to differentiate between facultative or obligate parasites and, within that last category, between holoparasitic or hemiparasitic weeds. The facultative parasites can be managed in similar ways to non-parasitic weeds, but they should be targeted early, before the plants start parasitizing the crop. The obligate parasitic weeds can be controlled through suicidal germination, or inhibition of seed germination and seedling attachment. This can be achieved by (inter- or rotation) crops or soil microbes producing germination stimulating or germination/attachment inhibiting compounds or by application of synthetic analogues of these compounds. Perhaps the most challenging component of parasitic weed management in smallholder farming systems is, however, to identify solutions that are locally available and match farm resources and capacities, and then to raise awareness among smallholders about the existence and application of these solutions.

DOI: 10.1079/9781789247657.0013

13.1 Introduction

As explained in Chapter 12, this volume, one of the main reasons parasitic weeds cause problems in smallholder cropping systems in Africa is the limited knowledge and application of effective control strategies by smallholder farmers. For some parasitic weed species there is limited knowledge on their biology and control. However, for many important species a range of effective control technologies is available, but farmers may lack awareness of these solutions or be unable to implement them because of the solution-specific requirements.

In this chapter, management of parasitic weeds will be discussed in relation to the effectiveness and relevance for African smallholder farming systems. The assessment of effectiveness depends on the degree to which the measure: (i) reduces existing seed banks; (ii) prevents further seed production; (iii) avoids seed dissemination; and (iv) reduces or prevents crop damage (Table 13.1). The relevance for smallholders in Africa is mainly determined by the availability, affordability and accessibility of the management strategy and whether it is compatible with the local farming and cropping system and their environments.

Seven broad categories are considered: sanitation, mechanical, chemical, cultural, biological, genetic and integrated control. Mechanical control is physical removal of the parasitic weed from its host, including hand weeding. Chemical control is the use of herbicides, or growth regulators, and different application techniques. With cultural control, we refer to measures based on agronomic interventions related to crop choices and combinations, as well as arrangements and timing of crop establishment and soil fertility or crop nutrition management. Biological control includes microbes (i.e. fungi and bacteria) and other biological control agents, such as seed predators, herbivores and bioherbicides (e.g. based on botanicals). Genetic control refers to the use of resistant or tolerant cultivars as well as herbicide-tolerant cultivars. In the last category, we discuss integrated control approaches, in which two or more practices from the other six categories are synergistically combined.

Following the order of treatments in previous sections of this book, the current chapter will first discuss stem parasites (including the annual *Cuscuta* spp. and perennial parasites such as species of the genera *Viscum*, *Tapinanthus*, *Erianthemum* and *Cassytha*), and then root parasites (including annual parasitic plant species of the genera *Striga*, *Rhamphicarpa*, *Alectra* and *Orobanche*/*Phelipanche*).

A search of the scientific literature database of Web of Science shows that control or management of witchweed (*Striga* spp.), broomrape (*Orobanche* spp. or *Phelipanche* spp.), mistletoe (*Viscum* spp., *Tapinanthus* spp., *Erianthemum* spp. or *Phragmanthera* spp.) and dodder (*Cuscuta* spp.) are the four most frequently studied groups of parasitic weeds (Fig. 13.1). Expressed as the share of all studies on a specific parasitic weed group, weed management is shown to be more frequently studied for witchweed and broomrape than mistletoe and dodder.

13.2 Management of Stem Parasites

There are several methods to manage stem parasites but a combination of two or more methods in an integrated management approach would achieve sufficient and long-lasting control (e.g. Lanini and Kogan, 2005).

Table 13.1. Overview of management options for parasitic weeds in cropping systems in Africa, with indications of effectiveness and relevance for smallholders.

	Effectiveness in reducing or preventing:					
Category	Seed bank	Seed production	Seed dissemination	Crop damage	Smallholder relevance	Downsides
Sanitation	+	+	++	+/–	+	Requires awareness, vigilance, discipline and time
Mechanical	+	++	–	+/–	+	Laborious, requires accurate timing
Chemical	+	++	+/–	+/–	–	Costly Limited availability Knowledge intensive Environmental and health issues
Cultural	++	+	+/–	+/–	++	Longer-term effectiveness Requires fundamental change Organic amendments carry the risk of introducing *Striga* spp. seed
Biological	+	+	+/–	+	+/–	Knowledge intensive, difficult to manage May require a commercial producer/operator/promoter
Genetic	+	+	+/–	++	+	Trade-offs with other desired cultivar traits Requires developed seed systems
Integrated	++	++	++	++	+/–	Costly and laborious

Highly effective/relevant (++), effective/relevant (+), moderately effective/relevant (+/–), not effective/relevant (–).

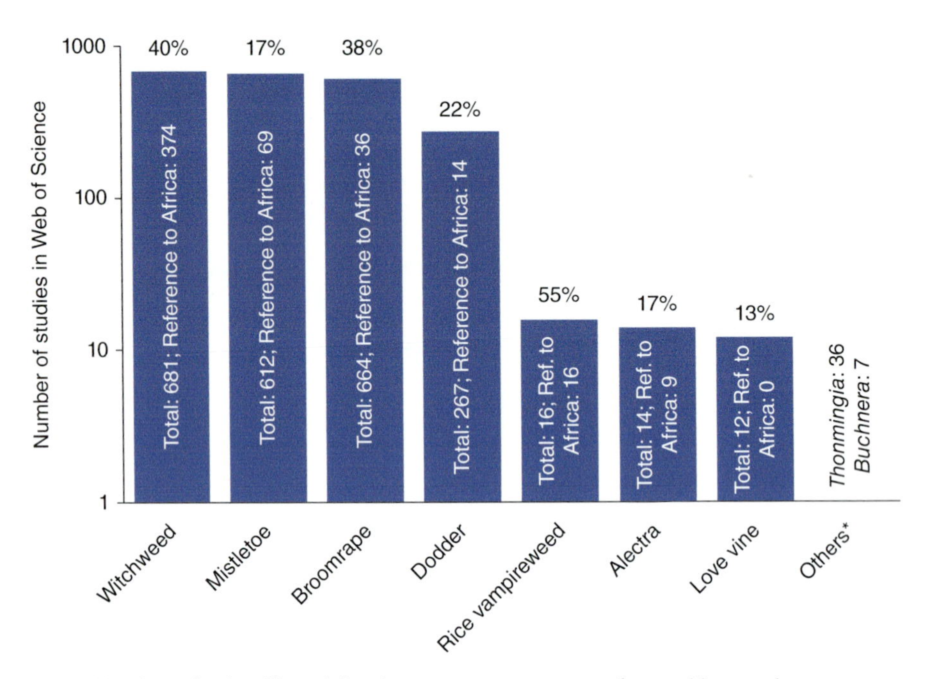

Fig. 13.1. Number of scientific publications on management of parasitic weeds, per group, as obtained from Web of Science (accessed 23 March 2021), and the number of these studies conducted with explicit reference to Africa or one of the African countries. Percentages above each bar represent the share of studies on management relative to the total number of studies on a specific parasitic weed management group. *Others: 36 studies on *Thonningia* and seven on *Buchnera hispida*, but none about their management.

There is a distinction between stem parasites such as dodder (*Cuscuta* spp.) that germinate in soil before attaching to a host and stem parasites that germinate directly on the host stem, such as love vine (*Cassytha* spp.) and mistletoe. The first group of parasites has a wider range of potential pre-attachment control measures than the second group. Dodder can be managed by targeting prevention of contamination, germination and seedling growth before attachments to a host (e.g. use of clean seed, tillage, soil-applied herbicide, cultural control measures aimed at out-shading), in addition to some post-infection measures. The second group of stem parasites is mostly controlled after infection (e.g. by pruning infected branches or applying systemic herbicides) and only to a limited extent by preventing new infections (e.g. by bird scaring, removing the parasite from nearby trees before flowering or the use of resistant tree species).

Obviously, prevention of parasitic infection would be preferred because the infection and post-infection interventions almost always have negative effects on the crop (e.g. Parker, 1991). In addition, post-infection measures rarely control the parasite.

13.2.1 Sanitation

Effective management approaches begins with measures preventing infestation of stem parasites. This requires an assessment of the main contamination and dispersal factors and vectors, such as domesticated or wild animals, agricultural equipment, crop seeds, and shoes or tyres (e.g. Rubiales and Fernández-Aparicio, 2012). For dodders, for example, the use of contaminated crop seeds is an important source of infestations (Parker and Riches, 1993). Dodder seeds in animal droppings are still viable even, to some extent, after manure composting (Yaacoby *et al.*, 2015), and therefore free-browsing cattle may be a source of contamination. For mistletoes and love vine, specific frugivorous bird species disperse fruits from infected trees to new trees (De Buen and Ornelas, 1999) and this means of contamination could be reduced by bird scaring. However, the feasibility, effectiveness and even ethics of bird scaring is questionable. It requires a lot of time and it may be difficult to sacrifice this amount of time just for a preventive measure. Also, in African agriculture, bird scaring is mostly done by labourers, including children (e.g. Bass, 2004), which prevents them from undertaking other economic activities or attending school.

Another way to avoid infection is by growing crop species that are not susceptible to the parasite. In dodder-infested fields, farmers could shift to non-host cereals (e.g. Dawson *et al.*, 1994). Growing a non-host crop for several years can lead to important reductions in the dodder seed bank, but, given the longevity of the seeds, infestation levels may quickly replenish after shifting back to a suitable host.

Love vine is an increasing problem in cashew production throughout Tanzania, but sources of contamination or factors causing its spread are largely unknown and therefore farmers use no sanitation and prevention measures (Kidunda *et al.*, 2017). Sanitation would normally start with careful surveying and timely monitoring of parasite infestations on cashew. Introduction of love vine seeds from parasite-infested to parasite-free farms could then be restricted by sanitation regulations in parasite-free farms. However, there are not many sanitary measures farmers can take, other than bird scaring and timely pruning of infected tree branches.

13.2.2 Mechanical control

Physical removal or pruning is one of the most effective control measures against stem parasites, and sometimes the only option, but obviously this is only relevant for infested trees; in annual crops, physical removal without causing concomitant damage to the crop plants themselves could prove rather difficult.

For the African smallholder the physical removal of dodders may be the only control measure in tree crops. A survey among coffee farmers in Uganda showed that physical removal (by hand) was the most popular method; the only other management option known to farmers was the application of herbicides, but this was only rarely conducted (Kagezi *et al.*, 2021). Physical removal requires the crop to be carefully and frequently monitored to eliminate young dodder plants as soon as they appear. Dodder that has been removed should immediately be burned.

The seed and germination biology of *Cuscuta* spp. provides some entry points for other effective mechanical control practices; such alternatives are mostly relevant for field crops where removal of the parasite from the crop canopy may be less feasible. First, germination and emergence of the parasite is mainly restricted to seeds found in the upper 1–2 cm of soil (e.g. Sandler *et al.*, 1997); second, the parasite seedlings lack roots. These features imply that shallow early tillage, before or at the start of crop season, or deeper tillage after the crop season, could be effective. However, an additional germination trait of *Cuscuta* spp. complicates this. Dodder seed requires scarification. Soil disturbance by tillage is one of the most obvious mechanisms of scarifying. Tillage therefore very likely increases germination rates and should be repeated several times to be an effective control measure. Additionally, the tillage should be in the short period between germination and host attachment, requiring careful monitoring. As soon as the parasite has established a connection with the host, the dodder seedling continues its growth independently from the soil (Parker and Riches, 1993) and tillage is no longer effective. Also, the risk with early shallow tillage is that the seedlings can regrow if left on the soil surface or only superficially buried, as they are quite tolerant to desiccation (Dawson *et al.*, 1994). Deep tillage could bury newly produced seeds to deeper soil layers from which they cannot germinate. This is another solution, but this is mostly suitable only for mechanized farms because doing this manually is labour intensive.

A last mechanical management option for *Cuscuta* spp. and other stem parasites is to remove the inflorescence before seed formation starts (between 50 and 60 days after first attachment). Again, this is a very time-consuming and laborious task, and therefore an infeasible control for farmers, in particular as damage to the crop cannot be avoided. Avoiding seed production and shedding is, however, crucial as dodder seeds are very long-lived (10–30 years, depending on species) (Lanini and Kogan, 2005). This would also mean that at harvest the crop plants must be cut as short as possible, to avoid surviving dodder attachments at the stem continuing their growth until seed production (Dawson *et al.*, 1994).

Control options for African mistletoe in cocoa or rubber are mostly limited to physical removal and tree pruning (Mathiasen *et al.*, 2008). This implies a trade-off between mistletoe control and tree production and survival; without any management, tree growth may be much impaired, but physical removal and tree pruning will also lead to production losses. The extent of pruning that would be necessary to remove the complete haustorial system (ranging from epicortical roots along the outside of tree branches to bark strands within host tree bark) may lead to severe impairment of tree performance. Control targeted at the parasite alone, by removing the aerial shoots of mistletoe, may lead to resprouting in post-control stages and therefore offers only temporary control. Resprouting of mistletoe could be partially prevented by wrapping black plastic around the infection point (Parker and Riches, 1993).

As love vine is exclusively a problem in tree crops (mainly mango and cashew), the mechanical control of love vine is not different to mistletoe control. Control should focus on early physical removal. The parasite can be removed from tree branches by hand or using a hooked stick, or infected branches can be pruned (Kidunda *et al.*, 2017).

13.2.3 Chemical control

Dodder can be effectively and selectively controlled by herbicides before it attaches to the host, but once it attaches the parasite becomes part of the host and very hard to kill selectively. Knowledge of the timing of crop and dodder growth stages, and a regular inspection of the crop for parasite presence, are both of crucial importance.

In the early stages of a dodder infestation, a contact herbicide can be applied for the prevention of any further spread (Rubiales and Fernández-Aparicio, 2012). A range of herbicide products have been identified for the control of dodder (e.g. Parker and Riches, 1993; Dawson *et al.*, 1994; Nadler-Hassar *et al.*, 2009) but these solutions are mostly relevant for mechanized agriculture in temperate regions, and the applicability (i.e. effectiveness, feasibility and availability) and therefore relevance for smallholder farmers in Africa remains to be ascertained.

Several herbicide solutions under tropical or subtropical conditions have provided good control (Table 13.2). Hock *et al.* (2008) observed good control of *C. campestris* with glyphosate (at 140 g ha^{-1}) under tropical conditions, but the effective and safe use of glyphosate would require adherence to strict protocols because of the broad-spectrum (non-selective) nature of this herbicide, implying it could also kill or damage the crop itself. Sulfosulfuron (at 50–100 g ha^{-1}), a herbicide that inhibits the enzyme acetolactate synthase (ALS), disrupting the synthesis of essential amino acids, has proven effective against *C. campestris* in tomato (Goldwasser *et al.*, 2012a), and application of slow-release granular formulations of pendimethalin has been a good selective method to effectively control this parasite in chickpea, as it germinates and emerges before it attaches to the crop plants (Goldwasser *et al.*, 2019). Pendimethalin alone or mixed with imazaquin applied pre-emergence, or pendimethalin alone applied post-emergence (at 10 days after sowing), effectively controlled dodder in niger (*Guizotia abyssinica;* Mishra *et al.*, 2007) and is effective in onion (Rao and Rao, 1993) and a range of other crops less commonly grown in Africa such as linseed (Mahere *et al.*, 2000) and lucerne (Dawson, 1990).

Table 13.2. Overview of proven herbicide effectiveness against field dodder (*Cuscuta campestris*), restricted to examples tested under subtropical to tropical climates.

Crop	Herbicide	Application	Source
Non-crop ornamental plants	Glyphosate	Post-emergence, spray applied[a]	Hock *et al.*, 2008
Tomato	Sulfosulfuron	Post-emergence, spray applied	Goldwasser *et al.*, 2012a; Eizenberg *et al.*, 2003
Chickpea	Pendimethalin	Granular, soil-applied	Goldwasser *et al.*, 2019
Niger, onion, linseed, lucerne	Pendimethalin	Post-emergence	Mishra *et al.*, 2007
Niger	Pendimethalin and imazaquin	Pre-emergence	Mishra *et al.*, 2007

[a]May cause injury to the host plant if not glyphosate resistant.

Exogenous application of compounds inhibiting the cell-wall-degrading enzymes, such as green tea catechins, were also found to provide effective control against dodder (Lewis *et al.*, 2010), but this has not been explored further as a control option in field crops.

In addition, as a sanitary measure, any chemical control of 'ordinary' weeds indirectly also contributes to the control of *Cuscuta* spp., as many weed species could act as a secondary host from which the parasite could spread to infest the crop. Glyphosate or 2,4-D (2,4-dichlorophenoxyacetic acid) could be used after crop harvest to kill surviving dodder plants and thereby prevent seed production (Dawson *et al.*, 1994).

Mistletoe is very difficult to control durably by chemical means, because of the long lifespan of their hosts (trees and perennial shrubs) allowing the parasite enough time to recover through resprouting from cortical strands or other reproductive structures. Re-infection through additional seed deposition by birds also remains likely. Systemic chemical control of an Australian mistletoe, *Amyema* spp., by trunk injection of glyphosate or a herbicide that combines aminopyralid and triclopyr, has proven partially effective in *Eucalyptus* spp., in particular for trees with greater than 30 cm stem diameter (Minko and Fagg, 1989), but effectiveness and risks with other mistletoe and host species would need to be carefully tested before application of this method. Also, this practice is laborious and requires injection equipment, the right product, and knowledge regarding rates and concentrations.

Spot application of the growth regulator ethephon could provide control of mistletoe, but the parasite may regenerate after the application and ethephon could also damage the host tree so application should be done with care. In addition, like many of the herbicide solutions, ethephon may not be reliably available in rural areas in Africa.

There is no known chemical control option available in Africa effective against love vine (Kidunda *et al.*, 2017).

13.2.4 Cultural control

Cuscuta spp. may germinate and grow in the absence of a host but usually seedlings are short-lived (<8 days). This enables management practices such as delayed crop sowing or transplanting, which have been shown to be effective against field dodder (*C. campestris*) in a niger crop in India (Mishra *et al.*, 2007). An additional advantage of transplanting is that older host plants may have higher resistance to *Cuscuta* spp., and a denser crop canopy can also suppress parasite seed germination or growth (Lanini and Kogan, 2005). The reason is that dodder is highly sensitive to shade. This implies that any other cultural method providing shade could be an additional control option. Using a non-host plant as a cover crop or intercrop to shade out the weed would be the most obvious way to achieve this. The complication with this option is the wide host range of dodder, including many forage legume species that would otherwise be excellent cover or intercrops. Another way to potentially increase dodder control through shading would be to increase the crop density or select cultivars with weed-suppressive traits such as early vigour and high leaf area index. To our knowledge, no attempts have been made to explore these potential control methods in field crops yet.

Keeping crops free of general weeds (both dicots and monocots, except for grasses; see Ashton and Santana, 1976) that can act as alternative hosts for *Cuscuta* spp. is a good additional step. Crop rotations offer another cultural control measure, if economically viable non-host crops are identified that are adapted to both the environmental and farming conditions. Growing cereal crops for several seasons, provided they are well weeded, would be suitable as a non-host rotation option and can contribute to parasite seed bank reductions (Dawson, 1987; Lanini, 2004). As dodder seeds have great longevity and parasitic weed seed banks are generally rapidly replenished when suitable hosts are grown again after a period of non-host growth, such rotations would need to be continued.

Seeds require scarification and water for imbibition, and these requirements offer another way of control. A stale seedbed approach could be followed, whereby a seedbed is prepared by primary and secondary tillage, followed by irrigation. Doing this a week before planting could cause suicidal germination of dodder seeds. To our knowledge, this approach has not been tested under field conditions. Lastly, postharvest dodder control would be effective to prevent replenishment of the seed bank by newly produced and viable dodder seeds. This can be done by burning crop residues (Orloff and Cudney, 1987) or applying glyphosate after crop harvest.

Love vine may be partially controlled by intercropping trees with shade-providing cover crops as these may suppress the growth of newly developing seedlings (suggested by Kidunda *et al.*, 2017). Mistletoe growth may also be suppressed by shade. Shade trees have been observed in Ghana to manage mistletoe in cocoa trees (Room, 1973). The principle of out-shading mistletoe could also be applied by stimulating the vigour and growth of the host tree itself, for instance by improved fertilization (Parker and Riches, 1993). The feasibility and economic viability of that approach for African farming systems is doubtful, however.

Medical and ethnobotanical studies note that stem parasites have medicinal (Kienle *et al.*, 2003; Ogunmefun *et al.*, 2013) or ornamental uses (Dzerefos *et al.*, 1999) and may therefore have economic value. Mistletoes are purposely grown in some areas (Ogunmefun *et al.*, 2013). In Tanzania, farmers may allow some trees to be infested by love vine, in particular trees near their homestead, for medicinal use (Kidunda *et al.*, 2017). Hence, if host damage from stem parasites such as mistletoe or love vine is acceptable, or if economic benefits from production of parasite-based medicines or ornamentals are greater than the combined value of the loss in tree products and the costs of control measures, smallholder farmers may decide to go for a low-intensity management approach aiming at the right balance between host and parasite production.

13.2.5 Biological control

Biological control for dodders has proven to be limited, both in terms of effectiveness and feasibility. The most widely used agent is the fungus *Colletotrichum gloeosporioides*. Isolates of *C. gloeosporioides* from dodder in soybean have some potential for control in field crops (Wang, 1986). *Fusarium oxysporum*, also effective against *Striga* spp., may control dodder before it attaches to a host, which requires timely and targeted applications (El-Dabaa *et al*., 2022). Kannan *et al*. (2014) tested isolates of *Trichoderma viride* and *Pseudomonas fluorescens* on seed and foliage to induce systemic resistance to *Cuscuta campestris* in chickpea, but this only led to a delay in parasite development.

The fungi *Geotrichum candidum* and *Alternaria alternata* have been identified as potential biocontrol agents against field dodder (Bewick *et al*., 1987). A bioherbicide based on *Alternaria destruens*, called Smolder™ (Sylvan Bioproducts, Inc.), was developed for the control of dodder in fruit trees/shrubs, such as cranberry and citrus in the USA, and was proven particularly effective in combination with glyphosate, ammonium sulfate and oil (Cook *et al*., 2009). Production of Smolder™ was not economically viable, however, and has been discontinued (M. Wach, Sylvan Inc., personal communication). Effectiveness of *A. destruens* against a range of dodder species encountered in Africa needs to be confirmed.

Water-dissolved extracts of some weed species, including some common weeds of African field crops such as *Cynodon dactylon* and *Sorghum halepense*, have provided field dodder control in lucerne, although high concentrations led to some concomitant damage to the crop itself (Habib and Rahman, 1988). Due to a lack of commercial viability this has not led to development and commercialization of plant-based bioherbicides, but such insights could provide leads for the future development of a component of integrated management based on locally sourced inputs that could help African smallholder farmers.

No biological control agents are known for the control of love vine or mistletoe.

13.2.6 Genetic control

The identification or development of resistant host genotypes of food crops against *Cuscuta* spp. has been suboptimal; perhaps the only two exceptions being resistance against *C. reflexa* observed in tomato (Kaiser *et al*., 2015) and resistance against *C. campestris* identified in Israelian chickpea genotypes ICCV 95333 and Hazera 4 (Goldwasser *et al*., 2012b). Genetic resistance against mistletoe has been observed in trees (see Mathiasen *et al*., 2008) and it is technically possible to develop mistletoe-resistant varieties of crop trees but, to our knowledge, this has not been done.

13.3 Management of Root Parasites

As noted earlier, important root-parasitic weeds in annual field crops in Africa are species of *Orobanche* (*O. crenata*, *O. minor*, and *O. cernua/cumana*), *Phelipanche* (*P. aegyptiaca*, *P. ramosa*), *Striga* (*S. asiatica*, *S. aspera*, *S. hermonthica* and *S. gesnerioides*), *Alectra* (*A. vogelii*, *A. sessiliflora* and *A. orobanchoides*) and *Rhamphicarpa fistulosa* (Parker, 2013). The facultative hemiparasite *Buchnera hispida* is sporadically observed but only locally recorded as an important weed problem (Gworgwor *et al.*, 2001).

13.3.1 Sanitation

Sanitation should be the first line of defence against root-parasitic weeds. Preventing parasitic weed seeds from entering a new arable field would require an understanding of seed dispersal factors and pathways. Seeds of root-parasitic weeds are tiny and can be distributed by a range of vectors, as shown with *Striga hermonthica* (Berner *et al.*, 1994a). Wind and water are the two environmental vectors that should be considered. Planting and maintaining vegetative windbreaks (i.e. hedges) around fields could prevent export or import of seeds through wind. In already contaminated fields, the permanent coverage of soils by living or dead biomass (i.e. cover crops or mulch) could avoid seeds being wind-blown or water-transported from these fields into neighbouring weed-free fields. Water runoff can also be slowed down or avoided by building barriers in or around fields (e.g. tied ridges, stone or soil bunds, planting pits or ditches). Browsing animals and agricultural activities, such as sowing and tillage, have been identified as the main factors of dispersal of *S. hermonthica* seeds (Berner *et al.*, 1994a), and these findings could be extended to other root-parasitic weeds as they all have similar seed properties. Farmers could deter roaming animals from entering by building fences or growing hedges around their fields. Field equipment used on a field infested with parasitic weed should be cleaned after operations, to avoid infestations of other fields thereafter (Goldwasser and Rodenburg, 2013). Indeed, sanitation of farm equipment was a major thrust of the successful United States Department of Agriculture (USDA) witchweed control programme in North and South Carolina (Eplee, 1992). Farmers should be made aware of the various pathways for parasitic seed dispersal and find ways to identify and interrupt them on their farms. This requires both vigilance and persistence in the application of sanitation measures.

In addition to the above, all of the management practices discussed below that contribute to reduction or prevention of parasitic weed seed production and seed bank replenishment could be considered as sanitation measures.

13.3.2 Mechanical control

All the above species can be controlled mechanically, by cutting and/or uprooting, using either mechanized (fuel- or cattle-driven) or hand-operated tillage devices for general weed control purposes, if crop damage by such interventions can be prevented. In smallholder farms in Africa, probably the most widespread means of mechanical parasitic weed control in annual crops is by short-handled hand hoes or by hand (Aflakpui *et al.*, 2008; Mrema *et al.*, 2017; N'cho *et al.*, 2019).

These control efforts are most effective and meaningful for facultative parasitic weed species, such as *Rhamphicarpa fistulosa* or *Buchnera hispida*, which can be removed before or shortly after finding a host. Timely weeding of these species will prevent parasitism and crop losses and contribute to a depletion of the soil seed bank as germinated seeds will not survive to produce seed.

However, for obligate parasitic weeds this management practice is ineffective in mitigating yield losses. Most of the damage caused by these root-parasitic weeds is inflicted during the invisible belowground stage of their development; mechanical control is only possible (without damaging the crop) when the parasites have emerged aboveground. For farmers dealing with obligate root-parasitic weeds in annual crops, it is still advisable to kill or remove these weeds, once aboveground, to prevent them from flowering and producing seeds. This recommendation would need to be followed up in postharvest stages as well. For example, for *Striga hermonthica* in sorghum, the remaining parasites are still able to reproduce during the first weeks after crop removal and therefore contribute to a replenishment of the soil seed bank (Rodenburg *et al.*, 2006b). By applying mechanical parasitic weed control, whether by hand or using implements, farmers can avoid or reduce parasitic weed seed production, with resultant lower infestation levels and reduced infection and crop damage in future cropping seasons. Farmers with access to machinery could deep-plough their field after or well before a crop season, as this would bury seeds to deeper soil layers where they could not germinate. This may be effective for species that require light for seed germination, such as *R. fistulosa*. For *Striga* spp. and *Orobanche* spp., with seeds that require being in the rooting zone to receive cues for their germination, this may be much less effective.

One of the main downsides of mechanical control is the risk of seed dissemination. Tillage operations may unintentionally aid the spread of parasitic weed seeds across a field (see Section 13.3.1 on sanitation). An additional, and perhaps more significant, disadvantage of mechanical control is the high energy and time requirement for tillage or hand-weeding operations. For instance, in smallholder rice systems in Africa, estimated hand-weeding times ranged between 173 (for one weeding operation) and 376 (for three weeding operations) person-hours ha^{-1} (Ogwuike *et al.*, 2014). In Africa, hand weeding is often done by women and children, negatively impacting general household welfare, childcare and schooling rates. The alternatives, such as fuel- or cattle-driven mechanized weeding implements, may save labour but will require availability and maintenance of machinery, additional expenses for fodder or fuel and possibly adaptations to crop arrangements to avoid crop damage during interventions.

Another alternative is refraining from frequent soil tillage or not tilling the soil at all; known as minimum or zero tillage respectively. This will reduce the entry of parasitic weed seeds into the root zone of cropped soils (Randrianjafizanaka *et al.*, 2018) and was effective against *S. hermonthica* in maize and sorghum, when combined with deep planting (Van Delft *et al.*, 2000), and also against *O. crenata* in faba bean, in particular when combined with low-rate glyphosate application (López-Bellido *et al.*, 2009).

13.3.3 Chemical control

All root-parasitic weeds in annual cropping systems can, in principle, be controlled by herbicides. Like mechanical control, a timely application of herbicides can be effective against facultative parasitic weed species, killing plants before they become parasitic and thereby avoiding both parasite reproduction and crop losses. For obligate root parasites, the use of herbicides will primarily contribute to the first (i.e. killing plants before they produce and disseminate seeds), but be less helpful in terms of avoiding current yield losses. When farmers rely only on traditional herbicide technologies targeting weed plants near or above the soil surface, obligate parasites could still damage the crop plants during their belowground stages. This is particularly true for holoparasites such as *Orobanche* spp. and *Phelipanche* spp., but also for hemiparasites such as *Striga* spp. and *Alectra* spp. Herbicides that target the weeds' photosynthesis cannot work for the control of holoparasites and underground obligate hemiparasites, in the absence of leaf chlorophyll. Another consideration is that the parasite and host are connected through their vascular system allowing systemic herbicides to transfer from the parasite to the host. The herbicide should therefore not be broad spectrum, or the host-plant cultivar should have resistance against the herbicide.

Importantly, any herbicide technology should be affordable and accessible to smallholder farmers, even in rural parts of Africa, and the farmer should know how to apply the herbicide safely and effectively. A survey among smallholder rice farmers in 20 countries in Africa revealed several problems. More than 60% of herbicide products available on rural markets across Africa appeared unregistered and therefore not quality-checked and the range of available herbicide products was often very limited and included some controversial products that are already banned in many countries. The vast majority of farmers (77%–89%) did not consult a formal information source (e.g. extension service or product label) regarding directions for safe and correct application of the chemical (Rodenburg *et al.*, 2019).

A number of effective herbicides are identified for use in parasitic-weed-infected field crops in Africa (Table 13.3). The main chemical control technologies used against parasitic weeds in Africa are herbicides that are applied to the aboveground plants (foliar application), the soil or the crop seeds. The latter can be done by seed soaking or seed coating, requiring concentrations that are non-toxic to the crop or crop varieties with herbicide resistance. Effective control of *Striga* spp. has been observed with the use of imidazolinone-resistant maize seeds coated with imazapyr (Gressel, 2009). It requires very small amounts of the herbicide and prevents successful witchweed parasitism at an early stage as the parasite dies soon after attachment to the host root (Kanampiu *et al.*, 2002, 2003). This offers effective control against *Striga* spp. in maize, in years with favourable rain distribution. Insufficient rainfall at the early crop stages (germination and emergence) or very high rainfall in the period thereafter, both negatively affect control efficacy. This problem could be solved by using slow-release formulations (Kanampiu *et al.*, 2009). A public–private partnership between the International Maize and Wheat Improvement Center

Table 13.3. Overview of proven herbicide effectiveness against root-parasitic weeds.

Crop	Parasite	Herbicide	Application[a]	Source
Cowpea	*Alectra vogelii*	Imazaquin	Seed soaking	Berner *et al*., 1994b
	Striga gesnerioides	Imazaquin	Seed soaking	Berner *et al*., 1994b
Maize	*S. asiatica*	Imazapyr	Seed coating	Kanampiu *et al*., 2003
	S. hermonthica	Imazapyr, pyrithiobac	Seed coating	Abayo *et al*., 1998; Kanampiu *et al*., 2002, 2003
Rice	*Rhamphicarpa fistulosa*	2,4-D amine	Post-emergence	Gbèhounou and Assigbé, 2003; Ouédraogo *et al*., 2017
	R. fistulosa	Piperophos	Early post-emergence	Ouédraogo *et al*., 2017
Sorghum	*S. asiatica*	Imazapyr, metsulfuron-methyl	Seed coating	Tuinstra *et al*., 2009
	S. hermonthica	2,4-D amine	Seed soaking	Dembele *et al*., 2005
Tomato	*Phelipanche aegyptiaca*	Sulfosulfuron	Soil-applied pre-emergence and foliar applied followed by overhead irrigation	Eizenberg *et al*., 2004; Eizenberg and Goldwasser, 2018
		Imazapic	Foliar applied followed by overhead irrigation, or applied through drip irrigation	Eizenberg and Goldwasser, 2018
Potato	*P. aegyptiaca*	Rimsulfuron	Soil-applied pre-emergence	Goldwasser *et al*., 2001
Carrot	*P. aegyptiaca*	Glyphosate	Sequential, post-emergence application	Foy *et al*., 1989; Cochavi *et al*., 2015
	Orobanche crenata	Glyphosate	Post-emergence	Jacobsohn and Kelman, 1980
	O. cumana	Imazapic	Post-emergence	Jacobsohn *et al*., 1996
Faba bean	*O. crenata*	Imazethapyr	Pre-emergence	Garcia Torres and Lopez Granados, 1991
	O. crenata	Pronamide	Seed soaking	Jurado Exposito *et al*., 1997

[a]Seed coating/seed soaking, often with acetolactate synthase (ALS) herbicides, is done with herbicide-resistant cultivars.

(CIMMYT), the African Agricultural Technology Foundation (AATF) and the chemical company BASF developed the principle of imidazolinone-resistant maize seeds coated with imazapyr into a commercially available technology, called StrigAway™. In 2016 AATF reported that over 170 t of StrigAway™ seed found its way to smallholder farms in Kenya, Tanzania and Uganda through seven seed companies, but 4 years later an AATF-commissioned impact assessment reported that seed companies failed to produce the anticipated amounts of imidazolinone-resistant maize seed and experienced major problems in processing and distribution of treated seed (Egerton University, 2020). Six of the seven seed companies have now stopped production and commercialization of StrigaAway™, leading to reduced availability of the technology in rural markets, except for Uganda.

For other crops, similar solutions have been investigated, but not developed into products. Seed soaking of cowpea in 1.8 mg ml^{-1} of imazaquin solution was proposed as a low-cost solution for smallholders to control *Striga gesnerioides* and *Alectra vogelii* (Berner *et al.*, 1994b). With sorghum seeds soaked for 5 min in a low concentration of 2,4-D, *S. hermonthica* infections were reduced (Dembele *et al.*, 2005). Unfortunately, this was not effective with millet.

For effective and safe chemical control of *O. aegyptiaca* in field-grown processing tomatoes, a decision-support system called PICKIT was developed for Israeli growers (Eizenberg and Goldwasser, 2018). PICKIT is based on a thermal time model predicting the development of parasitism and advises the grower when and how to employ a range of selective herbicides. Such innovations could also aid African farmers, provided they have access to information technology tools, good-quality herbicides and application equipment that enables the safe and efficient use of such chemicals.

Many systemic herbicides (e.g. glyphosate; imidazolinone herbicides such as imazapic, imazethapyr and imazamox; and sulfonylurea herbicides such as sulfosulfuron and rimsulfuron) have proven effective against *Orobanche* spp. and *Phelipanche* spp. (Eizenberg *et al.*, 2012; Eizenberg and Goldwasser, 2018) but, apart from glyphosate, availability of these products at rural agrochemical supply shops in Africa is likely to be erratic. The use of soil fumigation with chemicals such as methyl bromide or 1,3-dichloropropene, effective against *Orobanche* spp. and *Phelipanche* spp., is discouraged based on environmental concerns and costs (Eizenberg *et al.*, 2013) and not currently used by smallholders in Africa. Use of methyl bromide has been banned since 2005.

Another approach to chemical control of root-parasitic weeds is application of compounds that stimulate the germination of preconditioned seeds, leading to suicidal germination. This has been tested with the strigolactone analogues GR24 (Johnson *et al.*, 1976) and Nijmegen-1 (Zwanenburg and Thuring, 1997) but effectiveness in the field has not been great because of the instability of these compounds. Furthermore, the costs could become prohibitively high for sustained application in smallholder farmers' fields in Africa. Recently, new attempts have been made to develop a parasitic weed control technology based on this principle. Alongside Nijmegen-1, analogues derived from methyl phenlactonoates (i.e. MP1, MP3 and MP16) have been tested in millet and sorghum fields in Burkina Faso (Kountche *et al.*, 2019; Jamil *et al.*, 2022). These tests have reduced *Striga hermonthica* emergence by 60%–65% and could over time contribute to reduced parasitic weed seed banks (Jamil *et al.*, 2022). However, these studies indicated variability in effectiveness between fields and soil types, with better performance on sandy soils than on clayey soils (Kountche *et al.*, 2019), so ideally such technology should be part of an integrated approach. The main hurdle to actual implementation would be the current high cost of production of these analogues, as well as the lack of a reliable supply chain involving a local commercial partner, quality control and functioning markets.

13.3.4 Cultural control

Of all the available management solutions for parasitic weeds, cultural control options probably have the greatest potential to serve affected smallholder farmers because these options are not highly dependent on external inputs, markets, institutions or specialist knowledge. In addition, most of the practices also address additional crop production constraints such as drought, degraded or degrading soils, and competition from ordinary weeds. Under cultural control, we classify all decisions with respect to crop species, sequence, timing and spatial arrangement of establishment, but also fertilizer and water management. Time and space are the most important variables. Examples of cultural control measures include crop rotations, intercropping, the use of cover crops, mulching, organic or inorganic fertilizers, transplanting, deep planting, delayed or early planting and irrigation.

Soil fertility management, in the form of organic or inorganic fertilizers, is one of the most frequently suggested parasitic weed control measures, in particular against *Striga* spp. An early report by Porteres (1948) suggested that parasitic weeds of the Orobanchaceae are indicator species of infertile soils. One of the mechanistic explanations is that nitrogen- and phosphorus-deficient growing conditions stimulate the production and exudation of strigolactones by host-plant roots (Yoneyama *et al.*, 2007). Strigolactones comprise a category of plant hormones with a strong stimulating effect on parasitic weed seed germination. The reverse also seems to be true, as many studies have shown that nitrogen and phosphorus fertilizers contribute to a reduction of strigolactone production and consequently reduced *Striga* spp. infections in a range of host plants (Cechin and Press, 1993; Jamil *et al.*, 2011a). This is particularly clear under controlled experimental conditions. Suppressive effects of fertilizer application on parasitic weed infections have been observed in the field as well, but seem to be less consistent (Bebawi, 1981; Jamil *et al.*, 2012b; Tippe *et al.*, 2020), with some reports of insignificant effects (Kamara *et al.*, 2007; Tesso and Ejeta, 2011) and even some reports of a contradictory, stimulating effect at certain nitrogen application levels (Showemimo *et al.*, 2002). Against parasitic weed species of the genera *Phelipanche* and *Orobanche*, despite reductions in germination (Van Hezewijk and Verkleij, 1996) and seedling radicles (Westwood and Foy, 1999) of ammonium-exposed parasites in *in vitro* tests, applications of organic or inorganic fertilizers do not seem to help much in terms of broomrape control in the field (Goldwasser and Rodenburg, 2013).

In Israel, crops are seasonally fertilized but broomrape infection levels remain high (Y. Goldwasser, personal communication).

Farmers with high levels of parasitic weed infestation sometimes decide to abandon their field or part of the field and move to better areas (Houngbedji *et al.*, 2014; N'cho *et al.*, 2014). Such shifting cultivation practice is a common traditional strategy for farmers in Africa to deal with declining soil fertility and increased pest or disease pressure. After a long period (>10 years) of fallow, allowing restoration of soil fertility and reduction of pests or diseases, farmers may return to these fields. For parasitic weeds, this may only be effective if the fallow period is long enough, because many species of parasitic weeds have long-lived seeds. This strategy is therefore only feasible if the farmer has enough land. With increased land degradation and increasing human population densities, the practice of shifting cultivation has become less common (Demont *et al.*, 2007).

An alternative is a short-term improved fallow or a crop rotation, whereby the infested land is planted with a non-host species. Many annual root parasites have a specific host range and cannot parasitize all crop species. Parasites are often specialized for either dicotyledon or monocotyledon hosts; knowing this, the farmer can design a crop rotation that at least interrupts the seasonal cycle of parasitic weed reproduction and replenishment or build-up of the soil seed bank. For instance, farmers struggling with *Striga hermonthica* or *S. asiatica,* which parasitize cereal crops or sugarcane, can rotate the cereal crop (e.g. sorghum or millet) with an adapted non-cereal crop (e.g. groundnut, cowpea or sesame). Crop rotations or improved fallows can also be done with so-called false host species as trap crops. A wide range of crops that have been identified as non-hosts for a particular parasitic weed species can still cause seed germination of these parasites. These germinated parasites are unable to establish a viable connection with these false hosts and therefore die within a matter of days. This is called 'suicidal germination'. When these species are included in the improved fallow or rotation, the suicidal germination reduces the parasitic weed seed bank. Known trap crops for *Striga* spp. that parasitize cereals (e.g. *S. asiatica, S. aspera* and *S. hermonthica*) are groundnut, cowpea, pigeon pea, cotton, soybean and chickpea (e.g. Carson, 1989; Carsky *et al.*, 1994; Oswald and Ransom, 2001; Murdoch and Kunjo, 2003). For improved short fallows, leguminous cover crop species that can be used as trap crops against these *Striga* spp. are for instance *Aeschynomene histrix* (Merkel *et al.*, 2000), *Crotalaria ochroleuca* (Riches *et al.*, 2005) or *Stylosanthes guianensis* (Randrianjafizanaka *et al.*, 2018).

Cereals, on the other hand, can be used as trap crops for parasitic weed species parasitizing dicots. Seeds of *Phelipanche aegyptiaca* can germinate when exposed to root exudates of a wide number of non-host plant species, including wheat, sorghum and maize (Fernández-Aparicio *et al.*, 2007). False hosts of *Orobanche minor* include wheat (Lins *et al.*, 2006), oat and maize (Fernández-Aparicio *et al.*, 2007); sorghum and pearl millet induce germination of *O. cumana/cernua* (Fernández-Aparicio *et al.*, 2007). All these cereal species could be trap crops against broomrape. Sorghum, chickpea and African yam bean could be used as trap crops against *Striga gesnerioides* (Berner and Williams, 1998), and presumably at least sorghum could be used as a trap crop for *Alectra vogelii*. In a study in six Western African countries, Cardwell and Lane (1995) observed no *S. gesnerioides* in cowpea that was rotated with cotton or intercropped with rice; these could be useful cropping-system designs in areas infested by *S. gesnerioides*.

Some of these trap crops could also be intercropped, reducing the soil seed bank, as observed with legume intercropping in *Striga hermonthica*-infested cereal crops (Oswald *et al.*, 2002; Van Mourik *et al.*, 2008). If these trap crops are promiscuous, nitrogen-fixing legumes and mycotrophic, they may serve three simultaneous functions in the medium to long term: (i) stimulating suicidal germination of witchweed seeds; (ii) improving soil nitrogen; and (iii) enhancing the colonization of arbuscular mycorrhizal (AM) fungi that may improve the crop's phosphorus uptake as well as reducing levels of witchweed infestation (Lendzemo *et al.*, 2006). Trap crops that combine such triple roles with good environmental adaptation would need to be identified, and good-quality seed of such trap crops would then need to be made available to smallholders.

Intercropping or rotations can also be done with cover crops that are primarily grown for pest management or soil-improvement purposes. A prominent example of this is the use of desmodium (*Desmodium* spp.) in so-called push–pull systems (Figs 13.2, 13.3 and 13.4). Roots of desmodium exude compounds that stimulate *Striga* spp. germination, and allelopathic compounds inhibiting *Striga* spp. attachment, resulting in effective suicidal germination (Pickett *et al.*, 2010; Khan *et al.*, 2016). The terms 'push' and 'pull' refer to desmodium also being a repellent to stemborer insects and grass species such as napier grass (*Pennisetum purpureum*), grown in field margins, attracting these pest insects.

Other examples of witchweed-suppressive intercrop species are velvetleaf (*Mucuna* spp.) and stylo (*Stylosanthes guianensis*) (Fig. 13.3; Randrianjafizanaka *et al.*, 2018). These leguminous crops, whether used in rotation, as intercrop or as cover crop contribute to reductions in witchweed infection levels and seed production by causing suicidal germination. They also improve the soil nitrogen levels in the long term and reduce parasite seed dissemination when their living (as cover crop) or dead (as crop residue mulch) biomass sufficiently

Fig. 13.2. (A and B) Farmers showing their maize crop intercropped with stemborer-repellent and striga-suppressive silverleaf desmodium (*Desmodium uncinatum*) in Western Kenya. Napier grass (*Pennisetum purpureum*), grown to attract stemborers, is shown in the forefront.

Fig. 13.3. Two examples of effective striga-suppressive legume species that can be intercropped or rotated with cereals. (A) Silverleaf desmodium (*Desmodium uncinatum*) intercropped in maize; (B) desmodium flower; (C) stylo (*Stylosanthes guianensis*) intercropped with rice; (D) stylo flower.

Fig. 13.4. Sorghum grown in two adjacent *Striga hermonthica*-infested fields with (A) and without (B) desmodium intercrop.

protects the soil from runoff (Randrianjafizanaka *et al.*, 2018; Rodenburg *et al.*, 2020). Intercropping broomrape-susceptible crops with berseem clover (*Trifolium alexandrinum*) or oat (*Avena sativa*) has shown beneficial reduction of *O. crenata* infections due to allelopathic compounds inhibiting parasite germination (Fernández-Aparicio *et al.*, 2007, 2010).

The above-described crop rotation and intercrop solutions may be less relevant on soils containing seeds of more than one species of parasitic weed.

Under these conditions the control of one species may trade off with another species. For example, in central Malawi and central Tanzania maize and sorghum are grown on soils that are infested with *Striga asiatica* and *Alectra vogelii* (Kabambe *et al.*, 2008; J. Kayeke, personal communication). If farmers intercropped or rotated their cereal crop with legumes, to control *S. asiatica*, they could unintentionally maintain or increase the seed bank of *A. vogelii*. Resistant legume varieties, or non-hosts such as pigeon pea would be good candidates for cropping systems under such conditions (Kabambe *et al.*, 2008).

Changes in the timing of crop establishment could be another tool for parasitic weed control. Delayed sowing has proven effective to avoid infection by *Orobanche crenata* in faba bean and chickpea (Rubiales *et al.* 2003; Pérez-de-Luque *et al.* 2004; Grenz *et al.*, 2005), *Striga asiatica* in rainfed upland rice (Tippe *et al.*, 2017), *S. hermonthica* in maize and sorghum (Gbèhounou *et al.*, 2004) and *S. gesnerioides* in cowpea (Touré *et al.*, 1996). Delaying the crop establishment under rainfed arable conditions may negatively affect crop duration, grain filling time and concomitant crop yields (e.g. Gbèhounou *et al.*, 2004; Rubiales and Fernández-Aparicio, 2012). Farmers therefore need to seek an optimized timing for the specific conditions of their farm. This is a challenging endeavour given the increasingly unreliable rainfall patterns associated with climate change. In contrast with obligate parasites, the facultative parasitic weed *Rhamphicarpa fistulosa* benefited from a delayed rice crop establishment in rainfed lowlands (Tippe *et al.*, 2017), and consequently farmers growing rice in *R. fistulosa*-infested areas would be advised to plant as early as possible.

Crop transplanting from a fertilized nursery free of parasitic weeds may be helpful for rainfed lowland rice to delay infection times and reduce crop damage by *R. fistulosa*, but this would require additional labour. It has been suggested as a witchweed management option on free-draining soils as well, for instance with sorghum or maize (Oswald and Ransom, 2002; Van Ast *et al.*, 2005; Vissoh *et al.*, 2008), but under arid conditions the crop may suffer from a high transplanting shock, leading to potential crop losses.

13.3.5 Biological control

Phytophagous insects can be a biological control agent against parasitic weeds, but only if they are species specific, meaning they only predate on seeds or tissue of the parasite and not the host. Weevils of the genus *Smicronyx* (Curculionidae, Curculioninae, Smicronychini) may be potential witchweed biocontrol agents. The species *Smicronyx guineanus* is observed to predate on witchweed seeds in Africa (Anderson and Cox, 1997), whereas other *Smicronyx* species predate on *Cuscuta* spp. and *Orobanche* spp. (Caldara *et al.*, 2014). This could potentially be used as a biocontrol measure, but only when combined with other effective management options, because the large seed bank requires the weevil to have an extraordinarily high seed-destruction rate to be effective on its own (Smith *et al.*, 1993).

The larvae of another insect species, the fly *Phytomyza orobanchia* (Diptera: Agromyzidae) provides control of *Phelipanche ramosa* and *Orobanche cumana/cernua* in field-grown tomato in Ethiopia (Elzein *et al.*, 1999) and control of *O. crenata* in Morocco (Klein, 1995). The larvae of *P. orobanchia* predate on young *Orobanche* spp. seed capsules and can reduce seed production by 11%–90% (Sauerborn, 1991). Effective and long-term control of *Orobanche* spp. with this species requires an inundated approach, with the biocontrol agent mass produced and repeatedly released into the field, at a rate of more than 1000 adults ha^{-1} (under Moroccan conditions) for several subsequent years (Klein and Kroschel, 2002). Mass production of the biocontrol insects requires knowledge of insect ecology and biology as well as methodologies and facilities (e.g. a so-called phytomyzarium, a construction where parasitic weed shoots with pupae can be collected and grown into adults). The feasibility of this for smallholder farmers in Africa is questionable.

The observations and insights on the predation behaviour and damage of these phytophagous insects have, to our knowledge, not led to a coordinated control programme against parasitic weeds in smallholder farms in Africa. The feasibility and preconditions for an effective and viable control programme would require further study before it could be developed. Research and development focusing on this biocontrol approach have not advanced beyond investigations on insect ecology and interactions with parasitic weeds, and it is doubtful this can be a tangible control technology for smallholder farmers in Africa. There are nevertheless some things these farmers could do to at least stimulate or maintain good levels of natural infestations of phytophagous insects targeting parasitic weed species in their crop. They would need to refrain from the use of generic insecticides and deep tillage and even crop rotation and irrigation (Klein and Kroschel, 2002), although the latter two may not be advisable from an agronomic perspective. Weeded parasitic weed shoots could be collected and stored in the field margins to facilitate and enhance the rate of pupae survival. Farmers could maintain floristically diverse field margins, although empirical evidence for the effectiveness of these measures is lacking. For effective parasitic weed control, the natural population of the phytophagous insects would need to be strengthened (Klein and Kroschel, 2002). In addition, it is strongly advised to combine the use of insect-based biocontrol approaches with other parasitic weed management practices, such as intercropping with trap crops. This was an effective combination in *O. crenata*-infested broad bean fields in Egypt (Abu-Shall and Ragheb, 2014).

For the control of parasitic weeds, the use of bacteria or fungi, or compounds isolated from them, such as phytotoxins or natural amino acids such as tyrosine, leucine and methionine (e.g. Vurro *et al.*, 2009), as biological agents has been the subject of three decades of scientific studies. Microbes associated with the soil or host-plant rhizosphere may directly interfere with the host–parasite interaction. Such microbes can: (i) be pathogenic to the parasite; (ii) produce compounds that are antagonistic to the parasite or interfere with the biochemical-based communications between host and parasite; (iii) facilitate and increase nitrogen and phosphorus uptake by the host plant; (iv) alter the host-plant root architecture or exudate production; or (v) cause induced systemic resistance (Masteling *et al.*, 2019).

The most extensively studied biological control option is the use of phytopathogenic fungi, primarily of the genus *Fusarium* (e.g. *F. oxysporum*), as a so-called mycoherbicide. This could be used to coat crop seeds, which guarantees the establishment of the biocontrol agent in the rooting zone (and hence infection) at a minimum inoculum input rate (Elzein *et al.*, 2006). The fungus is generally pathogenic to *Striga* spp. but also appears an effective strigolactone-degrading agent (Boari *et al.*, 2016).

One of the weaknesses of any biological control approach is the applicability in smallholder farming systems, particularly in rural Africa. The technology would require inoculum of the control agent (e.g. the fungi), which depends on local production facilities, seed systems, and functional markets, and knowledge and equipment to pursue the recommended application techniques (e.g. seed coating). Setting up such a product value chain and training extension agents and farmers is not easy in remote areas with limited infrastructure and institutions. A potential solution has recently been developed and tested whereby toothpicks are infected by primary inoculum of *F. oxysporum* f. sp. *strigea*, FOXY T14, that overproduces amino acids (Nzioki *et al.*, 2016). These infected toothpicks can be used by farmers to prepare secondary field inoculum by sticking them in boiled rice for a few days. A small bottle-cap (half a teaspoon) of this inoculum is then applied to each crop planting hole. Field tests of this technology in 500 witchweed-infested farmers' fields in Kenya have shown maize yield to increase by more than 40% (Nzioki *et al.*, 2016). The toothpick technology has recently been made commercially available to smallholders in Kenya as Kichawi Kill™, but we have not seen any published data yet on adoption by smallholder cereal farmers.

Another biological control route for smallholder farmers would be to stimulate AM fungi residing in the soil (Lendzemo *et al.*, 2006). Host roots colonized by these fungi stop producing strigolactones that stimulate *Striga* spp. germination (López-Ráez *et al.*, 2011) and the host plants also become more efficient at acquiring phosphorus and nitrogen (De Boer *et al.*, 2005). Overall, AM fungi-colonized crop plants show reduced parasite damage compared with non-colonized host plants (Lendzemo and Kuyper, 2001). The colonization of AM fungi in arable soils can be stimulated by increasing the organic matter contents, as these correlate positively with AM fungi (Gaur and Adholeya, 2002), but under smallholder farming systems in Africa this may not be easily achieved because of the scarcity of, and competing claims on, plant biomass. Refraining from seasonal soil tillage (McGonigle *et al.*, 1990) and intercropping with mycotrophic crops (Lendzemo *et al.*, 2006) may also benefit AM fungi. Purposely inoculating witchweed-infested fields with AM fungi does not seem feasible and affordable for smallholder farmers, but if the conditions allow the crop to be transplanted from a smaller surface nursery, the nursery itself could be inoculated (Lendzemo *et al.*, 2006).

13.3.6 Genetic control

A range of cultivars has been identified with increased levels of resistance or tolerance against parasitic weeds (Table 13.4). Resistance is a category of genetically determined defence mechanisms that reduce parasite infection levels, whereas tolerance is the phenomenon of reduced negative effects from parasite infections (Rodenburg *et al.*, 2006a). Cultivars combining parasitic weed resistance or tolerance with broad environmental adaptation to African growing conditions would offer valuable and low-cost solutions to smallholders. Prominent examples of adapted resistant cultivars are observed in rice (NERICA varieties) and sorghum (e.g. Framida, SRN39, IS9830). On these cultivars, levels of infection by *Striga* spp. are generally much lower than on susceptible varieties of the same crop (Fig. 13.5). Although some resistance has been found in maize as well, there are fewer adapted cultivars with broad-spectrum resistance (Yacoubou *et al.*, 2021) currently available to smallholders in Africa.

Resistance against *S. gesnerioides* and *Alectra vogelii* has been observed in several cowpea accessions, but only a few of them combine this with other desirable agronomic or grain traits (e.g. Kamara *et al.*, 2008; Touré *et al.*, 1997). Exceptions are the *A. vogelii*-resistant varieties Komcalle, IT99K-573-2-1 and IT98K-205-8, which are already widely adopted by farmers in Burkina Faso (Dieni *et al.*, 2018), and *S. gesnerioides*-resistant Suvita-2 in Burkina and Mali (Touré *et al.*, 1997). Unfortunately, the latter cultivar was susceptible to *S. gesnerioides* ecotypes from Niger.

A wide range of faba bean varieties and breeding lines with resistance to *O. crenata* and *O. foetida* have been identified as well (Table 13.4). Many of these cultivars have resistance genes derived from the Egyptian variety Giza 402, an old variety developed by ICARDA, the International Center for Agricultural Research in the Dry Areas (Ter Borg *et al.*, 1994; Rubiales *et al.*, 2016). Some resistance has been found in chickpea, lentil and carrot (Table 13.4). Tomato cultivars with resistance against *Orobanche* spp. and *Phelipanche* spp. have been identified outside Africa (Qasem and Kasrawi, 1995; Hershenhorn *et al.*, 2009; Dor *et al.*, 2010; Tokasi *et al.*, 2014).

Across parasitic weed-resistant varieties, none of the identified materials was observed to be immune. With *S. hermonthica*, for instance, even very resistant varieties cannot completely prevent infection, and it takes the production of just 8–41 (depending on seed bank density) seed capsules per crop plant to replenish the estimated seed losses during a season (Rodenburg *et al.*, 2006b). Alectra seed capsules contain about five times more seed than *S. hermonthica*, and therefore the risk of seed bank replenishment or increases by newly produced seed is even more in *Alectra* spp. Individual parasitic plants that were able to bypass the resistance will not just contribute to seed bank replenishment, the seeds they produce also inherit their virulence and are therefore able to parasitize and reproduce in future crops despite resistance, leading to a completely new and virulent population. In addition to the risk of resistance breakdown in the long term, there is also a risk of crop damage in the short term. Resistance is not necessarily combined with tolerance, and even very resistant

Table 13.4. Cultivars with identified resistance or tolerance against annual root-parasitic weeds.

Parasite	Crop	Resistant	Tolerant	Source
Striga spp.	Maize	TZL COMP1 SYN; IWD C2 SYN; KSTP 94; STR-VE-216; H12 (0804-7STR); ZD05	Staha; H511; Maseno Double Cobber; SC535; SC527; MQ623; TZEEI-79, -74	Efron, 1993; Gurney *et al.*, 2002; Oswald and Ransom, 2004; Amusan *et al.*, 2008; Badu-Apraku and Oyekunle, 2012; Menkir *et al.*, 2012; Mutinda *et al.*, 2018; Nyakurwa *et al.*, 2018; Kamara *et al.*, 2020
	Sorghum	IS9830; Framida; SRN39; N13; Dobbs, 555	IS9830; Tiemarifing; Framida; Ochuti	El Hiweris, 1987; Gurney *et al.*, 1995; Haussmann *et al.*, 2000; Mohamed *et al.*, 2003; Rodenburg *et al.*, 2005; Rodenburg *et al.*, 2006a; Rodenburg *et al.*, 2008
	Millet	P 3908; Soowabya; ARD 80;	P 490; Dienidie; M141; M239; M029; M197; M017; KBH	Kountche *et al.*, 2013; Sattler *et al.*, 2018
	Rice	NERICA-1,-2, -3, -4, -5, -10; CG14; Umgar; M27; T2; B3913F-16-5-ST-42; IR49255-B-B-5-2; IR47255-B-B4; IR47697-4-3-1; FARO 40; Ble Chai; SCRID090-60-1-1-2-4; Ra-JL; 3293; Anakila; Super Basmati; Nipponbare	Makassa; CG14; ACC102196; NERICA-1, 10, -17; FARO 11; Botrabe	Harahap *et al.*, 1993; Johnson *et al.*, 1997; Gurney *et al.*, 2006; Rodenburg *et al.*, 2010, 2015, 2017; Cissoko *et al.*, 2011; Jamil *et al.*, 2011b, 2012a; Samejima *et al.*, 2016
	Cowpea	IT97K-499-35; IT90K-82-2; UAM09 1046-6-1; UAM09 1046-6-2; Suvita-2[a]; B301, IT82D-849		Kamara *et al.*, 2008; Omoigui *et al.*, 2017; Touré *et al.*, 1997
Alectra spp.	Cowpea	Komcalle; IT99K-573-2-1; IT98K-205-8		Dieni *et al.*, 2018

	Soybean	Bossier; Ocepara-4		Kabambe *et al.*, 2008
Rhamphicarpa fistulosa	Rice	Gambiaka; TOG5681; IR64; NERICA-L-23, -31, -40, -48	NERICA-L-39, -20; Supa India (Kilombero)	Rodenburg *et al.*, 2011 Rodenburg *et al.*, 2016
Orobanche spp.	Faba bean	Najeh; Chourouk; F 402 (Giza 402); 402/294; Baraca; Bader; XBJ90.03-16-1-1-1; XBJ90.04-6-2-1-1-4-C; XBJ90.04-2-3-1-1-1-2A; V-1268; V-1302; V-1301; V-268; V-231; V-319; V-1272; 674 / 154 / 85 L3-4; 402 / 29 / 84	Najeh; XAR-VF00.13-89-2-1-1-1-1; XBJ90.04-6-2–1-1-4-C	Ter Borg *et al.*, 1994; Abbes *et al.*, 2007, 2020; Rubiales *et al.*, 2014 Rubiales *et al.*, 2016; Trabelsi *et al.*, 2016; Amri *et al.*, 2019
	Chickpea	FLIP 98-22C; Nayer; X96TH62-A4-A1-W1-A1-A1-A1-A1		Nefzi *et al.*, 2016
	Lentil	LR9; V02		Mbasani-Mansi *et al.*, 2019
	Carrot	Palaiseau; Buror		Zehhar *et al.*, 2003

[a]Resistance is not broad spectrum as susceptibility against at least one parasite ecotype was observed.

Fig. 13.5. (A) Rice varieties resistant to *Striga hermonthica* adjacent to susceptible varieties in a variety-screening trial in Western Kenya. (B) Sorghum varieties resistant to *S. hermonthica* growing alongside susceptible varieties in variety-screening trials in Western Kenya (permission from Steven Runo). (C) Rice varieties resistant to *Rhamphicarpa fistulosa* (dark red-coloured plants) alongside susceptible varieties in a variety-screening trial in Tanzania. Lines indicate borders between different varieties.

crop varieties may therefore incur serious yield reductions, as shown with genotypic responses of sorghum to *S. hermonthica* (Rodenburg *et al.*, 2005). For these reasons (seed bank management and crop loss prevention), additional measures are still required. As a minimum, the use of resistant varieties should be combined with the (manual) removal of parasitic plants that have bypassed the resistance. Additional measures are preferably preventive rather than curative to avoid the need for additional inputs such as labour and agrochemicals. Preventive measures are ideally based on good agricultural practices and as such not only prevent parasitic weed infection and damage but also generally contribute to healthier and more productive crops and soils.

13.4 Integrated Management

Effective and durable control of parasitic weeds in annual cropping systems requires a long-term management programme combining locally adapted control measures. The importance of an integrated approach to parasitic weeds has long been acknowledged (e.g. Parker, 1991) but relatively few successful examples, in particular from Africa, have been reported.

Perhaps the most obvious and feasible integrated management approach is the combination of resistant and/or tolerant cultivars with any of the other control measures. A high-potential integrated management strategy entails combining resistant and/or tolerant cultivars with precision fertilization (i.e. a small amount of fertilizer with a tailored composition of macro- and micronutrients delivered to the host-plant seed, planting hole or foliage), but fine-tuning this still requires research and development (Mwangangi *et al.*, 2021).

The combination of resistant varieties and the biocontrol agent *Fusarium oxysporum* f. sp. *strigea* has also been suggested as a promising approach, but this would likewise require further research (Shayanowako *et al.*, 2018). Growing herbicide-resistant or witchweed-resistant maize varieties (and possibly varieties that combine these traits) with a legume intercrop, such as *Desmodium uncinatum*, results in sharply reduced *S. hermonthica* infections and good grain yields with the added advantage of long-term effects on reduced parasite seed banks (Kanampiu *et al.*, 2018). Other suggested integrated management concepts include combining no-till with crop residue mulching and trap crop intercropping (Randrianjafizanaka *et al.*, 2018); shallow tillage or no-till with (deep) transplanting (Van Ast *et al.*, 2005); or transplanting of seedlings from an AM-fungi-colonized nursery (Lendzemo *et al.*, 2006).

13.5 Concluding Remarks on Management of Parasitic Weeds

There are many technologies for effective management of parasitic weeds but very few of them are broadly used by smallholders in Africa. Perhaps the most important reason for the poor rate of adoption of these technologies is the lack of awareness among farmers in rural parts of Africa, with high illiteracy rates, ill-informed and poorly resourced agricultural extension services, and poor internet access (see Chapter 12, this volume). Another reason for scarce implementation of control technologies on smallholder farms in Africa is that few technological ideas have been further developed into commercially available and affordable solutions. There are some exceptions, such as the witchweed-resistant or tolerant sorghum (e.g. Framida, SRN39) and rice (e.g. NERICA-1 and -10) varieties, the imazapyr-resistant maize technology StrigAway™ and the more recently released Kichawi Kill™, but even for these technologies, high adoption levels among smallholder farmers across the continent have not yet been reached. Finding the underlying reasons for the slow development of effective and accessible technologies, despite the excellent scientific work being conducted worldwide, and for the disappointing adoption of those technologies that have ultimately made it to a market release (e.g. StrigAway™), would be a sensible first step to undertake. The example of the only seed company to continue with the production and distribution of StrigAway™ maize seed, in Uganda, should also be analysed and could serve as a successful model for other companies and technologies. Findings from such analyses of failures and successes could be used to inform the design of a new approach for parasitic weed technology and dissemination among smallholder farmers.

Given that most of the affected crops are staple food crops with relatively low monetary value, and most of the affected farmers are smallholders that often partly grow crops to satisfy the food demands of their immediate family with little capacity for investments in costly inputs, the incentives for commercial partners to invest in technology development and retail in Africa is largely lacking. Technologies that are highly dependent on industries and markets may, at least under the current conditions, not provide the most likely and sustainable control for farmers affected by parasitic weed in Africa.

On the other hand, solutions based on locally available resources or solutions that address multiple crop production constraints simultaneously without high additional costs, such as preventive management and cultural management solutions, may have a higher potential at present and for the immediate future. Examples are sanitation measures to prevent new infestations, frequent and early tillage for *Cuscuta* spp., deep tillage for facultative parasites such as *Rhamphicarpa fistulosa* or optimized crop establishment times. Particularly effective and suitable, at least for the root-parasitic weeds, are any practices that improve soil fertility and stimulate AM fungi and suicidal germination of parasitic weed seeds, for example improved crop rotations and intercropping, no-till and crop residue mulching and the use of locally sourced organic soil amendments. In addition to their local availability and low costs, the practices that focus on soil have a high adoption potential because they also improve crop performance and resilience against other pests and diseases. For such technologies to be widely accepted, farmers need to be aware of their existence and convinced of their effectiveness while understanding that the effects are mostly expressed in the medium to longer term. Agricultural extension services and other forms of information transfer, such as radio and video disseminated by local media and mobile cinema services, could be intensified to achieve this. Several videos on witchweed biology and management in sorghum, maize and rice have been developed in the past (see, for example, www.accessagriculture.org accessed 31 July 2023). Another approach is to incorporate biology and management of parasitic weeds as a topic in school curricula, as was done in witchweed-infested regions in Tanzania (Riches *et al.*, 2005). These information sources should focus first on parasitic weed biology, as this will be fundamental for making informed decisions in the field. Second, they should promote an integrated management approach against parasitic weeds, as this is likely to be most effective in the short and longer term.

References

Abayo, G.O., English, T., Eplee, R.E., Kanampiu, F.K., Ransom, J.K. *et al.* (1998) Control of parasitic witchweeds (*Striga* spp.) on corn (*Zea mays*) resistant to acetolactate synthase inhibitors. *Weed Science* 46, 459–466.

Abbes, Z., Kharrat, M., Delavault, P., Simier, P. and Chaibi, W. (2007) Field evaluation of the resistance of some faba bean (*Vicia faba* L.) genotypes to the parasitic weed *Orobanche foetida* Poiret. *Crop Protection* 26, 1777–1784.

Abbes, Z., Boualleglue, A., Trabelsi, I., Trabelsi, N., Taamalli, A. *et al.* (2020) Investigation of some biochemical mechanisms involved in the resistance of faba bean (*Vicia faba* L.) varieties to *Orobanche* spp. *Plant Protection Science* 56, 317–328.

Abu-Shall, A.M.H. and Ragheb, E.I.M. (2014) Management of *Orobanche crenata* using trap crops and *Phytomyza orobanchia* Kalt. in broad bean (*Vicia faba*) field in Egypt. *Egyptian Journal of Biological Pest Control* 24, 217–223.

Aflakpui, G.K.S., Bolfrey-Arku, G.E.K., Anchirinah, V.M., Manu-Aduening, J.A. and Adu-Tutu, K.O. (2008) Incidence and severity of *Striga* spp. in the coastal savanna zone of Ghana: results and implications of a formal survey. *Outlook on Agriculture* 37, 219–224.

Amri, M., Trabelsi, I., Abbes, Z. and Kharrat, M. (2019) Release of a new faba bean variety "Chourouk" resistant to the parasitic plants *Orobanche foetida* and *O. crenata* in Tunisia. *International Journal of Agriculture and Biology* 21, 499–505.

Amusan, I.O., Rich, P.J., Menkir, A., Housley, T. and Ejeta, G. (2008) Resistance to *Striga hermonthica* in a maize inbred line derived from *Zea diploperennis*. *New Phytologist* 178, 157–166.

Anderson, D.M. and Cox, M.L. (1997) *Smicronyx* species (Coleoptera: Curculionidae), economically important seed predators of witchweeds (*Striga* spp.) (Scrophulariaceae) in sub-Saharan Africa. *Bulletin of Entomological Research* 87, 3–17.

Ashton, F.M. and Santana, D. (1976) Cuscuta *Spp. (Dodder): A Literature Review of its Biology and Control*. Cooperative Extension Bulletin. University of California, Berkeley, California.

Badu-Apraku, B. and Oyekunle, M. (2012) Genetic analysis of grain yield and other traits of extra-early yellow maize inbreds and hybrid performance under contrasting environments. *Field Crop Research* 129, 99–110.

Bass, L.E. (2004) *Child Labor in Sub-Saharan Africa*. Lynne Rienner Publishers, Boulder, Colorado.

Bebawi, F.F. (1981) Response of sorghum cultivars and striga population to nitrogen fertilization. *Plant and Soil* 59, 261–267.

Berner, D.K. and Williams, O.A. (1998) Germination stimulation of *Striga gesnerioides* seeds by hosts and nonhosts. *Plant Disease* 82, 1242–1247.

Berner, D.K., Cardwell, K.F., Faturoti, B.O., Ikie, F.O. and Williams, O.A. (1994a) Relative roles of wind, crop seeds, and cattle in dispersal of *Striga* spp. *Plant Disease* 78, 402–406.

Berner, D.K., Awad, A.E. and Aigbokhan, E.I. (1994b) Potential of imazaquin seed treatment for control of *Striga gesnerioides* and *Alectra vogelii* in cowpea (*Vigna unguiculata*). *Plant Disease* 78, 18–23.

Bewick, T.A., Binning, L.K., Stevenson, W.R. and Stewart, J. (1987) A mycoherbicide for control of swamp dodder (*Cuscuta gronovii* Willd) Cuscutaceae. In: Weber, H.C. and Forstreuter, W. (eds) *Proceedings of the 4th International Symposium on Parasitic Flowering Plants. Marburg, Germany, 1987*. Philipps-Universität, Marburg, Germany, pp. 93–104.

Boari, A., Ciasca, B., Pineda-Martos, R., Lattanzio, V.M.T., Yoneyama, K. *et al.* (2016) Parasitic weed management by using strigolactone-degrading fungi. *Pest Management Science* 72, 2043–2047.

Caldara, R., Franz, N.M. and Oberprieler, R.G. (2014) Curculioninae Latreille, 1802. In: Leschen, R.A.B. and Beutel, R.G. (eds) *Handbook of Zoology. Coleoptera, Beetles – Morphology and Systematics*. Vol. 3. De Gruyter, Berlin, pp. 589–628.

Cardwell, K.F. and Lane, J.A. (1995) Effect of soils, cropping system and host phenotype on incidence and severity of *Striga gesnerioides* on cowpea in West Africa. *Agriculture, Ecosystems & Environment* 53, 253–262.

Carsky, R.J., Singh, L. and Ndikawa, R. (1994) Suppression of *Striga hermonthica* on sorghum using a cowpea intercrop. *Experimental Agriculture* 30, 349–358.

Carson, A.G. (1989) Effect of intercropping sorghum and groundnuts on density of *Striga hermonthica* in the Gambia. *Tropical Pest Management* 35, 130–132.

Cechin, I. and Press, M.C. (1993) Nitrogen relations of the sorghum *Striga hermonthica* host–parasite association: germination, attachment and early growth. *New Phytologist* 124, 681–687.

Cissoko, M., Boisnard, A., Rodenburg, J., Press, M.C. and Scholes, J.D. (2011) New Rice for Africa (NERICA) cultivars exhibit different levels of post-attachment resistance against the parasitic weeds *Striga hermonthica* and *Striga asiatica*. *New Phytologist* 192, 952–963.

Cochavi, A., Achdari, G., Smirnov, E., Rubin, B. and Eizenberg, H. (2015) Egyptian broomrape (*Phelipanche aegyptiaca*) management in carrot under field conditions. *Weed Technology* 29, 519–528.

Cook, J.C., Charudattan, R., Zimmerman, T.W., Rosskopf, E.N., Stall, W.M. *et al.* (2009) Effects of *Alternaria destruens*, glyphosate, and ammonium sulfate individually and integrated for control of dodder (*Cuscuta pentagona*). *Weed Technology* 23, 550–555.

Dawson, J.H. (1987) *Cuscuta* (Convolvulaceae) and its control. In: Weber, H.C. and Forstreuter, W. (eds) *Proceedings of the 4th International Symposium on Parasitic Flowering Plants, Marburg, Germany, 1987*. Philipps-Universität, Marburg, Germany, pp. 137–149.

Dawson, J.H. (1990) Dodder (*Cuscuta* spp.) control with dinitroaniline herbicides in alfalfa (*Medicago sativa*). *Weed Technology* 4, 341–348.

Dawson, J.H., Musselman, L.J., Wolswinkel, P. and Dörr, I. (1994) Biology and control of *Cuscuta*. *Reviews of Weed Science* 6, 265–317.

De Boer, W., Folman, L.B., Summerbell, R.C. and Boddy L. (2005) Living in a fungal world: impact of fungi on soil bacterial niche development. *FEMS Microbiology Reviews* 29, 795–811.

De Buen, L.L. and Ornelas, J.F. (1999) Frugivorous birds, host selection and the mistletoe *Psittacanthus schiedeanus*, in central Veracruz, Mexico. *Journal of Tropical Ecology* 15, 329–340.

Dembele, B., Dembele, D. and Westwood, J.H. (2005) Herbicide seed treatments for control of purple witchweed (*Striga hermonthica*) in sorghum and millet. *Weed Technology* 19, 629–635.

Demont, M., Jouve, P., Stessens, J. and Tollens, E. (2007) Boserup versus Malthus revisited: evolution of farming systems in northern Côte d'Ivoire. *Agricultural Systems* 93, 215–228.

Dieni, Z., Tignegre, J.B.D., Tongoona, P., Dzidzienyo, D., Asante, I.K. *et al.* (2018) Identification of sources of resistance to *Alectra vogelii* in cowpea *Vigna unguiculata* (L.) Walp. germplasm from Burkina Faso. *Euphytica* 214: 234.

Dor, E., Alperin, B., Wininger, S., Ben-Dor, B., Somvanshi, V.S. *et al.* (2010) Characterization of a novel tomato mutant resistant to the weedy parasites *Orobanche* and *Phelipanche* spp. *Euphytica* 171, 371–380.

Dzerefos, C., Shackleton, C. and Witkowski, E. (1999) Sustainable utilization of woodrose-producing mistletoes (Loranthaceae) in South Africa. *Economic Botany* 53, 439–477.

Efron, Y. (1993) Screening maize for tolerance to *Striga hermonthica*. *Plant Breeding* 110, 192–200.

Egerton University (2020) *Impact Evaluation of AATF's 'Striga Control in Maize Project'*. Tegemeo Institute of Agricultural Policy and Development, Egerton University, Nairobi.

Eizenberg, H. and Goldwasser, Y. (2018) Control of Egyptian broomrape in processing tomato: a summary of 20 years of research and successful implementation. *Plant Disease* 102, 1477–1488.

Eizenberg, H., Goldwasser, Y., Achdary, G. and Hershenhorn, J. (2003) The potential of sulfosulfuron to control troublesome weeds in tomato. *Weed Technology* 17, 133–137.

Eizenberg, H., Goldwasser, Y., Golan, S., Plakhine, D. and Hershenhorn, J. (2004) Egyptian broomrape (*Orobanche aegyptiaca*) control in tomato with sulfonylurea herbicides – greenhouse studies. *Weed Technology* 18, 490–496.

Eizenberg, H., Aly, R. and Cohen, Y. (2012) Technologies for smart chemical control of broomrape (*Orobanche* spp. and *Phelipanche* spp.). *Weed Science* 60, 316–323.

Eizenberg, H., Hershenhorn, J., Ephrath, J.H. and Kanampiu, F. (2013) Chemical control. In: Joel, D.M., Gressel, J. and Musselman, L.J. (eds) *Parasitic Orobanchaceae: Parasitic Mechanisms and Control Strategies*. Springer, Berlin, pp. 415–432.

El-Dabaa, M.A.T., Abo-Elwafa, G.A. and Abd-El-Khair, H. (2022) Safe methods as alternative approaches to chemical herbicides for controlling parasitic weeds associated with nutritional crops: a review. *Egyptian Journal of Chemistry* 65, 53–65.

El Hiweris, S.O. (1987) Nature of resistance to *Striga hermonthica* (Del.) Benth. parasitism in some *Sorghum vulgare* (Pers.) cultivars. *Weed Research* 27, 305–312.

Elzein, A.E.M., Kroschel, J., Admasu, A. and Fetene, M. (1999) Preliminary evaluation of *Phytomyza orobanchia* (Diptera: Agromyzidae) as a controller of *Orobanche* spp. in Ethiopia. *SINET: Ethiopian Journal of Science* 22 , 271–282.

Elzein, A.E.M., Kroschel, J. and Leth, V. (2006) Seed treatment technology: an attractive delivery system for controlling root parasitic weed *Striga* with mycoherbicide. *Biocontrol Science and Technology* 16, 3–26.

Eplee, R.E. (1992) Witchweed (*Striga asiatica*): an overview of management strategies in the USA. *Crop Protection* 11, 3–7.

Fernández-Aparicio, M., Sillero, J.C. and Rubiales, D. (2007) Intercropping with cereals reduces infection by *Orobanche crenata* in legumes. *Crop Protection* 26, 1166–1172.

Fernández-Aparicio, M., Emeran, A.A. and Rubiales, D. (2010) Inter-cropping with berseem clover (*Trifolium alexandrinum*) reduces infection by *Orobanche crenata* in legumes. *Crop Protection* 29, 867–871.

Foy, C.L., Jain, R. and Jacobsohn, R. (1989) Recent approaches for chemical control of broomrape (*Orobanche* spp.). *Reviews of Weed Science* 4, 123–152.

Garcia Torres, L. and Lopez Granados, F. (1991) Progress of herbicide control of broomrape (*Orobanche* spp.) in legumes and sunflower (*Helianthus annuus* L.). In: Ransom, J.K., Musselman, L.J., Worsham, A.D. and Parker, C. (eds) *Proceedings of the 5th International Symposium of Parasitic Weeds, 24–30 June 1991*. International Maize and Wheat Improvement Center (CIMMYT), Nairobi, pp. 306–309.

Gaur, A. and Adholeya, A. (2002) Arbuscular-mycorrhizal inoculation of five tropical fodder crops and inoculum production in marginal soil amended with organic matter. *Biology and Fertility of Soils* 35, 214–218.

Gbèhounou, G. and Assigbé, P. (2003) *Rhamphicarpa fistulosa* (Hochst.) Benth. (Scrophulariaceae): new pest on lowland rice in Benin. Results of a survey and immediate control possibilities. *Annales des Sciences Agronomique du Bénin* 4, 89–103.

Gbèhounou, G., Adango, E., Hinvi, J.C. and Nonfon, R. (2004) Sowing date or transplanting as components for integrated *Striga hermonthica* control in grain-cereal crops? *Crop Protection* 23, 379–386.

Goldwasser, Y. and Rodenburg, J. (2013) Integrated agronomic management of parasitic weed seed banks. In: Joel, D.M., Gressel, J. and Musselman, L.J. (eds) *Parasitic Orobanchaceae: Parasitic Mechanisms and Control Strategies*. Springer, Berlin, pp. 393–413.

Goldwasser, Y., Eizenberg, H., Hershenhorn, J., Plakhine, D., Blumenfeld, T. *et al.* (2001) Control of *Orobanche aegyptiaca* and *O. ramosa* in potato. *Crop Protection* 20, 403–410.

Goldwasser, Y., Mario, R., Sazo, M. and Lanini, W.T. (2012a) Control of field dodder (*Cuscuta campestris*) parasitizing tomato with ALS-inhibiting herbicides. *Weed Technology* 26, 740–746.

Goldwasser, Y., Miryamchik, H., Sibony, M. and Rubin, B. (2012b) Detection of resistant chickpea (*Cicer arietinum*) genotypes to *Cuscuta campestris* (field dodder). *Weed Research* 52, 122–130.

Goldwasser, Y., Rabinovitz, O. and Hayut, E. (2019) Selective and effective control of field dodder (*Cuscuta campestris*) in chickpea with granular pendimethalin. *Weed Technology* 33, 586–594.

Grenz, J.H., Manschadi, A.M., Uygurc, F.N. and Sauerborn, J. (2005) Effects of environment and sowing date on the competition between faba bean (*Vicia faba*) and the parasitic weed *Orobanche crenata*. *Field Crops Research* 93, 300–313.

Gressel, J. (2009) Crops with target-site herbicide resistance for *Orobanche* and *Striga* control. *Pest Management Science* 65, 560–565.

Gurney, A.L., Press, M.C. and Ransom, J.K. (1995) The parasitic angiosperm *Striga hermonthica* can reduce photosynthesis of its sorghum and maize hosts in the field. *Journal of Experimental Botany* 46, 1817–1823.

Gurney, A.L., Taylor, A., Mbwaga, A., Scholes, J.D. and Press, M.C. (2002) Do maize cultivars demonstrate tolerance to the parasitic weed *Striga asiatica*? *Weed Research* 42, 299–306.

Gurney, A.L., Slate, J., Press, M.C. and Scholes, J.D. (2006) A novel form of resistance in rice to the angiosperm parasite *Striga hermonthica*. *New Phytologist* 169, 199–208.

Gworgwor, N.A., Ndahi, W.B. and Weber, H.C. (2001) Parasitic weeds of north-eastern region of Nigeria: a new potential threat to crop production. *Proceedings of the BCPC conference, 13–15 November 2001, Brighton, UK*. British Crop Protection Council, Farnham, UK, pp. 181–186.

Habib, S.A. and Rahman, A. (1988) Evaluation of some weed extracts against field dodder on alfalfa (*Medicago sativa*). *Journal of Chemical Ecology* 14, 443–453.

Harahap, Z., Ampong-Nyarko, K. and Olela, J.C. (1993) *Striga hermonthica* resistance in upland rice. *Crop Protection* 12, 229–231.

Haussmann, B.I.G., Hess, D.E., Reddy, B.V.S., Welz, H.G. and Geiger, H.H. (2000) Analysis of resistance to *Striga hermonthica* in diallel crosses of sorghum. *Euphytica* 116, 33–40.

Hershenhorn, J., Eizenberg, H., Dor, E., Kapulnik, Y. and Goldwasser, Y. (2009) *Phelipanche aegyptiaca* management in tomato. *Weed Research* 49, 34–47.

Hock, S.M., Wiecko, G. and Knezevic, S.Z. (2008) Glyphosate dose affected control of field dodder (*Cuscuta campestris*) in the tropics. *Weed Technology* 22, 151–155.

Houngbedji, T., Pocanam, Y., Shykoff, J., Nicolardot, B. and Gibot-Leclerc, S. (2014) A new major parasitic plant in rice in Togo: *Rhamphicarpa fistulosa*. *Cahiers Agricultures* 23, 357–365.

Jacobsohn, R. and Kelman, Y. (1980) Effectiveness of glyphosate in broomrape (*Orobanche* spp.) control in four crops. *Weed Science* 28, 692–699.

Jacobsohn, R., Tanaami, Z. and Eizenberg, H. (1996) Selective control of broomrape in carrot and vetch with foliar-applied imidazolinone herbicides. *Phytoparasitica* 24: 207.

Jamil, M., Charnikhova, T., Cardoso, C., Jamil, T., Ueno, K. *et al.* (2011a) Quantification of the relationship between strigolactones and *Striga hermonthica* infection in rice under varying levels of nitrogen and phosphorus. *Weed Research* 51, 373–385.

Jamil, M., Rodenburg, J., Charnikhova, T. and Bouwmeester, H.J. (2011b) Pre-attachment *Striga hermonthica* resistance of New Rice for Africa (NERICA) cultivars based on low strigolactone production. *New Phytologist* 192, 964–975.

Jamil, M., Charnikhova, T., Houshyani, B., Van Ast, A. and Bouwmeester, H.J. (2012a) Genetic variation in strigolactone production and tillering in rice and its effect on *Striga hermonthica* infection. *Planta* 235, 473–484.

Jamil, M., Kanampiu, F.K., Karaya, H., Charnikhova, T. and Bouwmeester, H.J. (2012b) *Striga hermonthica* parasitism in maize in response to N and P fertilisers. *Field Crops Research* 134, 1–10.

Jamil, M., Wang, J.Y., Yonli, D., Ota, T., Berqdar, L., Traore, H. *et al.* (2022) *Striga hermonthica* suicidal germination activity of potent strigolactone analogs: evaluation from laboratory bioassays to field trials. *Plants* 11: 1045.

Johnson, A.W., Rosebery, G. and Parker, C. (1976) A novel approach to *Striga* and *Orobanche* control using synthetic germination stimulants. *Weed Research* 16, 223–227.

Johnson, D.E., Riches, C.R., Diallo, R. and Jones, M.J. (1997) Striga on rice in West Africa: crop host range and the potential of host resistance. *Crop Protection* 16, 153–157.

Jurado Exposito, M., Garcia Torres, L. and Castejon Munoz, M. (1997) Broad bean and lentil seed treatments with imidazolinones for the control of broomrape (*Orobanche crenata*). *Journal of Agricultural Science* 129, 307–314.

Kabambe, V., Katunga, L., Kapewa, T. and Ngwira, A.R. (2008) Screening legumes for integrated management of witchweeds (*Alectra vogelii* and *Striga asiatica*) in Malawi. *African Journal of Agricultural Research* 3, 708–715.

Kagezi, G.H., Kyalo, G., Kobusinge, J., Nkuutu, E., Baluku, J. *et al.* (2021) A rapid assessment of the invasive dodder weed, *Cuscuta* spp. on robusta coffee, *Coffea robusta* in Busoga coffee growing sub-region, Eastern Uganda. *East African Scholars Journal of Agriculture and Life Sciences* 4, 55–66.

Kaiser, B., Vogg, G., Fürst, U.B. and Albert, M. (2015) Parasitic plants of the genus *Cuscuta* and their interaction with susceptible and resistant host plants. *Frontiers in Plant Science* 6: 45.

Kamara, A.Y., Menkir, A., Chikoye, D., Omoigui, L.O. and Ekeleme, F. (2007) Cultivar and nitrogen fertilization effects on *Striga* infestation and grain yield of early maturing tropical maize. *Maydica* 52, 415–423.

Kamara, A.Y., Chikoye, D., Ekeleme, F., Omoigui, L.O. and Dugje, I.Y. (2008) Field performance of improved cowpea varieties under conditions of natural infestation by the parasitic weed *Striga gesnerioides*. *International Journal of Pest Management* 54, 189–195.

Kamara, A.Y., Menkir, A., Chikoye, D., Tofa, A.I., Fagge, A.A. *et al.* (2020) Mitigating *Striga hermonthica* parasitism and damage in maize using soybean rotation, nitrogen application, and *Striga*-resistant varieties in the Nigerian savannas. *Experimental Agriculture* 56, 620–632.

Kanampiu, F.K., Ransom, J.K., Friesen, D. and Gressel, J. (2002) Imazapyr and pyrithiobac movement in soil and from maize seed coats to control *Striga* in legume intercropping. *Crop Protection* 21, 611–619.

Kanampiu, F.K., Kabambe, V., Massawe, C., Jasi, L., Friesen, D. *et al.* (2003) Multi-site, multi-season field tests demonstrate that herbicide seed-coating herbicide-resistance maize controls *Striga* spp. and increases yields in several African countries. *Crop Protection* 22, 697–706.

Kanampiu, F., Karaya, H., Burnet, M. and Gressel, J. (2009) Needs for and effectiveness of slow release herbicide seed treatment *Striga* control formulations for protection against early season crop phytotoxicity. *Crop Protection* 28, 845–853.

Kanampiu, F., Makumbi, D., Mageto, E., Omanya, G., Waruingi, S. *et al.* (2018) Assessment of management options on *Striga* infestation and maize grain yield in Kenya. *Weed Science* 66, 516–524.

Kannan, C., Kumar, B., Aditi, P. and Gharde, Y. (2014) Effect of native *Trichoderma viride* and *Pseudomonas fluorescens* on the development of *Cuscuta campestris* on chickpea, *Cicer arietinum*. *Journal of Applied and Natural Science* 6, 844–851.

Khan, Z., Midega, C.A.O., Hooper, A. and Pickett, J. (2016) Push-pull: chemical ecology-based integrated pest management technology. *Journal of Chemical Ecology* 42, 689–697.

Kidunda, B.R., Kasuga, L.J. and Alex, G. (2017) Assessing the existence spread and control strategies of parasitic weed (*Cassytha filiformis*) on cashew trees in Tanzania. *Journal of Advanced Agricultural Technologies* 4, 285–289.

Kienle, G.S., Berrino, F., Büssing, A., Portalupi, E., Rosenzweig, S. *et al.* (2003) Mistletoe in cancer. A systematic review on controlled clinical trials. *European Journal of Medical Research* 8, 109–119.

Klein, O. (1995) Untersuchungen zur Populationsdynamik und zur Verwendung von Phytomyza orobanchia in der biologischen Bekämpfung von Orobanche spp. in Morocco. MSc thesis, Institute of Plant Production and Agroecology in the Tropics and Subtropics, University of Hohenheim, Stuttgart, Germany.

Klein, O. and Kroschel, J. (2002) Biological control of *Orobanche* spp. with *Phytomyza orobanchia*, a review. *BioControl* 47, 245–277.

Kountche, B.A., Hash, C.T., Dodo, H., Laoualy, O., Sanogo, M.D. *et al.* (2013) Development of a pearl millet *Striga*-resistant genepool: response to five cycles of recurrent selection under *Striga*-infested field conditions in West Africa. *Field Crops Research* 154, 82–90.

Kountche, B.A., Jamil, M., Yonli, D., Nikiema, M.P., Blanco-Ania, D. *et al.* (2019) Suicidal germination as a control strategy for *Striga hermonthica* (Benth.) in smallholder farms of sub-Saharan Africa. *Plants, People, Planet* 1, 107–118.

Lanini, W.T. (2004) Economical methods of controlling dodder in tomatoes. *Proceedings of the California Weed Science Society* 56, 57–59.

Lanini, W.T. and Kogan, M. (2005) Biology and management of *Cuscuta* in crops. *International Journal of Agriculture and Natural Resources* 32, 127–141.

Lendzemo, V.W. and Kuyper, T.W. (2001) Effects of arbuscular mycorrhizal fungi on damage by *Striga hermonthica* on two contrasting cultivars of sorghum, *Sorghum bicolor. Agriculture, Ecosystems & Environment* 87, 29–35.

Lendzemo, V.W., Van Ast, A. and Kuyper, T.W. (2006) Can arbuscular mycorrhizal fungi contribute to *Striga* management on cereals in Africa? *Outlook on Agriculture* 35, 307–311.

Lewis, K.C., Alers-Garciá, J. and Wright, L.J. (2010) Green tea catechins applied to susceptible hosts inhibit parasitic plant attachment success. *Crop Science* 50, 253–264.

Lins, R.D., Colquhoun, J.B. and Mallory-Smith, C.A. (2006) Investigation of wheat as a trap crop for control of *Orobanche minor*. *Weed Research* 46, 313–318.

López-Bellido, R.J., Benítez-Vega, J. and López-Bellido, L. (2009) No-tillage improves broomrape control with glyphosate in faba-bean. *Agronomy Journal* 101, 1394–1399.

López-Ráez, J.A., Charnikhova, T., Fernández, I., Bouwmeester, H. and Pozo, M.J. (2011) Arbuscular mycorrhizal symbiosis decreases strigolactone production in tomato. *Journal of Plant Physiology* 168, 294–297.

Mahere, J., Yadav, P.K. and Sharma, R.S. (2000) Chemical weed control in linseed with special reference to *Cuscuta*. *Indian Journal of Weed Science* 32, 216–217.

Masteling, R., Lombard, L., De Boer, W., Raaijmakers, J.M. and Dini-Andreote, F. (2019) Harnessing the microbiome to control plant parasitic weeds. *Current Opinion in Microbiology* 49, 26–33.

Mathiasen, R.L., Shaw, D.C., Nickrent, D.L. and Watson, D.M. (2008) Mistletoes. Pathology, systematics, ecology, and management. *Plant Disease* 92, 988–1006.

Mbasani-Mansi, J., Briache, F.Z., Ennami, M., Gaboun, F., Benbrahim, N. *et al.* (2019) Resistance of Moroccan lentil genotypes to *Orobanche crenata* infestation. *Journal of Crop Improvement* 33, 306–326.

McGonigle, T.P., Evans, D.G. and Miller, M.H. (1990) Effect of soil disturbance on mycorrhizal colonization and phosphorus absorption by maize in growth chamber and field experiments. *New Phytologist* 116, 629–636.

Menkir, A., Makumbi, D. and Franco, J. (2012) Assessment of reaction patterns of hybrids to *Striga hermonthica* (Del.) Benth. under artificial infestation in Kenya and Nigeria. *Crop Science* 52, 2528–2537.

Merkel, U., Peters, M., Tarawali, S.A., Schultze-Kraft, R. and Berner, D.K. (2000) Characterization of a collection of *Aeschynomene histrix* in subhumid Nigeria. *Journal of Agricultural Science* 134, 293–304.

Minko, G. and Fagg, P.C. (1989) Control of some mistletoe species on eucalypts by trunk injection with herbicides. *Australian Forestry* 52, 94–102.

Mishra, J.S., Moorthy, B.T.S., Bhan, M. and Yaduraju, N.T. (2007) Relative tolerance of rainy season crops to field dodder (*Cuscuta campestris*) and its management in niger (*Guizotia abyssinica*). *Crop Protection* 26, 625–629.

Mohamed, A., Ellicott, A., Housley, T.L. and Ejeta, G. (2003) Hypersensitive response to *Striga* infection in Sorghum. *Crop Science* 43, 1320–1324.

Mrema, E., Shimelis, H., Laing, M. and Bucheyeki, T. (2017) Farmers' perceptions of sorghum production constraints and *Striga* control practices in semi-arid areas of Tanzania. *International Journal of Pest Management* 63, 146–156.

Murdoch, A.J. and Kunjo, E.M. (2003) Depletion of natural soil seedbanks of *Striga hermonthica* in West Africa under different integrated management regimes. *Aspects of Applied Biology* 69, 261–268.

Mutinda, S.M., Masanga, J., Mutuku, J.M., Runo, S. and Alakonya, A. (2018) KSTP 94, an open-pollinated maize variety has post attachment resistance to purple witchweed (*Striga hermonthica*). *Weed Science* 66, 525–529.

Mwangangi, I.M., Büchi, L., Haefele, S., Bastiaans, L., Runo, S. *et al.* (2021) Combining host plant defence with targeted nutrition: key to durable control of hemiparasitic *Striga* in cereals in sub-Saharan Africa? *New Phytologist* 230, 2164–2178.

Nadler-Hassar, T., Shaner, D.L., Nissen, S., Westra, P. and Rubin, B. (2009) Are herbicide-resistant crops the answer to controlling *Cuscuta*? *Pest Management Science* 65, 811–816.

N'cho, S.A, Mourits, M., Rodenburg, J., Demont, M. and Lansink, A.O. (2014) Determinants of parasitic weed infestation in rainfed lowland rice in Benin. *Agricultural Systems* 130, 105–115.

N'cho, S.A., Mourits, M., Rodenburg, J. and Lansink, A.O. (2019) Inefficiency of manual weeding in rainfed rice systems affected by parasitic weeds. *Agricultural Economics* 50, 151–163.

Nefzi, F., Trabelsi, I., Amri, M., Triki, E., Kharrat, M. *et al.* (2016) Response of some chickpea (*Cicer arietinum* L.) genotypes to *Orobanche foetida* Poir. parasitism. *Chilean Journal of Agricultural Research* 76, 170–178.

Nyakurwa, C.S., Gasura, E., Setimela, P.S., Mabasa, S., Rugare, J.T. *et al.* (2018) Reaction of new quality protein maize genotypes to *Striga asiatica*. *Crop Science* 58, 1201–1218.

Nzioki, H.S., Oyosi, F., Morris, C.E., Kaya, E., Pilgeram, A.L. *et al.* (2016) *Striga* biocontrol on a toothpick: a readily deployable and inexpensive method for smallholder farmers. *Frontiers in Plant Science* 7: 1121.

Ogunmefun, O.T., Fasola, T.R., Saba, A.B. and Oridupa, O.A. (2013) The ethnobotanical, phytochemical and mineral analyses of *Phragmanthera incana* (Klotzsch), a species of mistletoe growing on three plant hosts in South-Western Nigeria. *International Journal of Biomedical Science* 9, 33–40.

Ogwuike, P., Rodenburg, J., Diagne, A., Agboh-Noameshie, A.R. and Amovin-Assagba, E. (2014) Weed management in upland rice in sub-Saharan Africa: impact on labor and crop productivity. *Food Security* 6, 327–337.

Omoigui, L.O., Kamara, A.Y., Ajeigbe, H.A., Akinwale, R.O., Timko, M.P. *et al.* (2017) Performance of cowpea varieties under *Striga gesnerioides* (Willd.) Vatke infestation using biplot analysis. *Euphytica* 213–244.

Orloff, S.B. and Cudney, D.W. (1987) Control of dodder in alfalfa with dinitroaniline herbicides. *Proceedings of the Western Society of Weed Science* 40, 98–103.

Oswald, A. and Ransom, J.K. (2001) *Striga* control and improved farm productivity using crop rotation. *Crop Protection* 20, 113–120.

Oswald, A. and Ransom, J.K. (2002) Response of maize varieties to transplanting in *Striga*-infested fields. *Weed Science* 50, 392–396.

Oswald, A. and Ransom, J.K. (2004) Response of maize varieties to *Striga* infestation. *Crop Protection* 23, 89–94.

Oswald, A., Ransom, J.K., Kroschel, J. and Sauerborn, J. (2002) Intercropping controls *Striga* in maize-based farming systems. *Crop Protection* 21, 367–374.

Ouédraogo, O., Kaboré, I., Kaboré, T. and Boussim, I.J. (2017) Effets de traitements herbicides sur *Rhamphicarpa fistulosa* (Hochst.) Benth. Une plante parasite facultative. *Journal of Applied Biosciences* 119, 11983–11992.

Parker, C. (1991) Protection of crops against parasitic weeds. *Crop Protection* 10, 6–22.

Parker, C. (2013) The parasitic weeds of the Orobanchaceae. In: Joel, D.M., Gressel, J. and Musselman, L.J. (eds) *Parasitic Orobanchaceae: Parasitic Mechanisms and Control Strategies.* Springer, Berlin, pp. 313–344.

Parker, C. and Riches, C.R. (1993) *Parasitic Weeds of the World: Biology and Control*. CAB International, Wallingford, UK.
Pérez-de-Luque, A., Sillero, J.C., Cubero, J.I. and Rubiales, D. (2004) Effect of sowing date and host resistance on the establishment and development of *Orobanche crenata* on faba bean and common vetch. *Weed Research* 44, 282–288.
Pickett, J.A., Hamilton, M.L., Hooper, A.M., Khan, Z.R. and Midega, C.A.O. (2010) Companion cropping to manage plants. *Annual Review of Phytopathology* 48, 161–177.
Porteres, R. (1948) Plants indicating the fertility level of the edapho-climatic cultural complex in tropical Africa. *Agronomie Tropicale* 3, 246–257.
Qasem, J.R. and Kasrawi, M.A. (1995) Variation of resistance to broomrape (*Orobanche ramosa*) in tomatoes. *Euphytica* 81, 109–114.
Randrianjafizanaka, M.T., Autfray, P., Andrianaivo, A.P., Ramonta, I.R. and Rodenburg, J. (2018) Combined effects of cover crops, mulch, zero-tillage and resistant varieties on *Striga asiatica* (L.) Kuntze in rice–maize rotation systems. *Agriculture, Ecosystems & Environment* 256, 23–33.
Rao, K.N. and Rao, R.S.N. (1993) Control of *Cuscuta* with herbicides in onion. In: *Proceedings of International Symposium on Integrated Weed Management for Sustainable Agriculture*. Indian Society of Weed Science, Hisar, India, pp. 196–198.
Riches, C.R., Mbwaga, A.M., Mbapila, J. and Ahmed, G.J.U. (2005) Improved weed management delivers increased productivity and farm incomes from rice in Bangladesh and Tanzania. *Aspects of Applied Biology* 75, 127–138.
Rodenburg, J., Bastiaans, L., Weltzien, E. and Hess, D.E. (2005) How can field selection for *Striga* resistance and tolerance in sorghum be improved? *Field Crop Research* 93, 34–50.
Rodenburg, J., Bastiaans, L. and Kropff, M.J. (2006a) Characterization of host tolerance to *Striga hermonthica*. *Euphytica* 147, 353–365.
Rodenburg, J., Bastiaans, L., Kropff, M.J. and Van Ast, A. (2006b) Effects of host plant genotype and seedbank density on *Striga* reproduction. *Weed Research* 46, 251–263.
Rodenburg, J., Bastiaans, L., Schapendonk, A.H.C.M., Van Der Putten, P.E.L., Van Ast, A. *et al.* (2008) CO_2-assimilation and chlorophyll fluorescence as indirect selection criteria for host tolerance against *Striga*. *Euphytica* 160, 75–87.
Rodenburg, J., Riches, C.R. and Kayeke, J.M. (2010) Addressing current and future problems of parasitic weeds in rice. *Crop Protection* 29, 210–221.
Rodenburg, J., Zossou-Kouderin, N., Gbèhounou, G., Ahanchede, A., Touré, A. *et al.* (2011) *Rhamphicarpa fistulosa*, a parasitic weed threatening rain-fed lowland rice production in sub-Saharan Africa – a case study from Benin. *Crop Protection* 30, 1306–1314.
Rodenburg, J., Cissoko, M., Kayeke, J., Dieng, I., Khan, Z.R. *et al.* (2015) Do NERICA rice cultivars express resistance to *Striga hermonthica* (Del.) Benth. and *Striga asiatica* (L.) Kuntze under field conditions? *Field Crop Research* 170, 83–94.
Rodenburg, J., Cissoko, M., Dieng, I., Kayeke, J. and Bastiaans, L. (2016) Rice yields under *Rhamphicarpa fistulosa*-infested field conditions, and variety selection criteria for resistance and tolerance. *Field Crop Research* 194, 21–30.
Rodenburg, J., Cissoko, M., Kayongo, N., Dieng, I., Bisikwa, J. *et al.* (2017) Genetic variation and host-parasite specificity of *Striga* resistance and tolerance in rice: the need for predictive breeding. *New Phytologist* 214, 1267–1280.
Rodenburg, J., Johnson, J.M., Dieng, I., Senthilkumar, K., Vandamme, E. *et al.* (2019) Status quo of chemical weed control in rice in sub-Saharan Africa. *Food Security* 11, 69–92.
Rodenburg, J., Randrianjafizanaka, M.T., Buchi, L., Dieng, I., Andrianaivo, A.P. *et al.* (2020) Mixed outcomes from conservation practices on soils and *Striga*-affected yields of a low-input, rice–maize system in Madagascar. *Agronomy for Sustainable Development* 40: 8.
Room, P.M. (1973) Ecology of the mistletoe *Tapinanthus bangwensis* growing on cocoa in Ghana. *Journal of Ecology* 61, 729–742.

Rubiales, D. and Fernández-Aparicio, M. (2012) Innovations in parasitic weeds management in legume crops. A review. *Agronomy for Sustainable Development* 32, 433–449.

Rubiales, D., Alcántara, C., Pérez-de-Luque, A., Gil, J. and Sillero, J.C. (2003) Infection of chickpea (*Cicer arietinum*) by crenate broomrape (*Orobanche crenata*) as influenced by sowing date and weather conditions. *Agronomie* 23, 359–362.

Rubiales, D., Flores, F., Emeran, A.A., Kharrat, M., Amri, M. *et al.* (2014) Identification and multi-environment validation of resistance against broomrapes (*Orobanche crenata* and *Orobanche foetida*) in faba bean (*Vicia faba*). *Field Crop Research* 166, 58–65.

Rubiales, D., Rojas-Molina, M.M. and Sillero, J.C. (2016) Characterization of resistance mechanisms in faba bean (*Vicia faba*) against broomrape species (*Orobanche* and *Phelipanche* spp.). *Frontiers in Plant Science* 7: 1747.

Samejima, H., Babiker, A.G., Mustafa, A. and Sugimoto, Y. (2016) Identification of *Striga hermonthica*-resistant upland rice varieties in Sudan and their resistance phenotypes. *Frontiers in Plant Science* 7: 634.

Sandler, H.A., Else, M.J. and Sutherland, M. (1997) Application of sand for inhibition of swamp dodder (*Cusucta gronovii*) seedling emergence and survival on cranberry (*Vaccinium macrocarpon*) bogs. *Weed Technology* 11, 318–323.

Sattler, F.T., Sanogo, M.D., Kassari, I.A., Angarawai, I.I., Gwadi, K.W. *et al.* (2018) Characterization of West and Central African accessions from a pearl millet reference collection for agro-morphological traits and *Striga* resistance. *Plant Genetic Resources* 16, 260–272.

Sauerborn, J. (1991) *Parasitic Flowering Plants: Ecology and Management.* Verlag Josef Margraf, Wiekersheim, Germany.

Shayanowako, A.T., Laing, M., Shimelis, H. and Mwadzingeni, L. (2018) Resistance breeding and biocontrol of *Striga asiatica* (L.) Kuntze in maize: a review. *Acta Agriculturae Scandinavica, Section B – Soil and Plant Science* 68, 110–120.

Showemimo, F.A., Kimbeng, C.A. and Alabi, S.O. (2002) Genotypic response of sorghum cultivars to nitrogen fertilization in the control of *Striga hermonthica*. *Crop Protection* 21, 867–870.

Smith, M.C., Holt, J. and Webb, M. (1993) Population model of the parasitic weed *Striga hermonthica* (Scrophulariaceae) to investigate the potential of *Smicronyx umbrinus* (Coleoptera: Curculionidae) for biological control in Mali. *Crop Protection* 12, 470–476.

Ter Borg, S.J., Willemsen, A., Khalil, S.A., Saber, H.A., Verkleij, J.A.C. *et al.* (1994) Field study of the interaction between *Orobanche crenata* Forsk. and some new lines of *Vicia faba* L. in Egypt. *Crop Protection* 13, 611–616.

Tesso, T.T. and Ejeta, G. (2011) Integrating multiple control options enhances *Striga* management and sorghum yield on heavily infested soils. *Agronomy Journal* 103, 1464–1471.

Tippe, D.E., Rodenburg, J., Van Ast, A., Anten, N.P.R., Dieng, I. *et al.* (2017) Delayed or early sowing: timing as parasitic weed control strategy in rice is species and ecosystem dependent. *Field Crops Research* 214, 14–24.

Tippe, D.E., Bastiaans, L., Van Ast, A., Dieng, A., Cissoko, M. *et al.* (2020) Fertilisers differentially affect facultative and obligate parasitic weeds of rice and only occasionally improve yields in infested fields. *Field Crops Research* 254: 107845.

Tokasi, S., Aval, M.B., Mashhadi, H.R. and Ghanbari, A. (2014) Screening of resistance to Egyptian broomrape infection in tomato varieties. *Planta Daninha* 32, 109–116.

Touré, M., Olivier, A., Ntare, B.R. and St-Pierre, C.A. (1996) The influence of sowing date and irrigation of cowpea on *Striga gesnerioides* emergence. In: Moreno, M.T., Cubero, J.I., Berner, D., Joel, D., Musselman L.J. and Parker, C. (eds) *Advances in Parasitic Plant Research. Proceedings of the Sixth International Symposium on Parasitic Weeds.* Dirección General de Investigación Agraria, Córdoba, Spain, pp. 451–455.

Touré, M., Olivier, A., Ntare, B.R., Lane, J.A. and St-Pierre, C.A. (1997) Inheritance of resistance to *Striga gesnerioides* biotypes from Mali and Niger in cowpea (*Vigna unguiculata* (L.) Walp.). *Euphytica* 94, 273–278.

Trabelsi, I., Abbes, Z., Amri, M. and Kharrat, M. (2016) Study of some resistance mechanisms to *Orobanche* spp. infestation in faba bean (*Vicia faba* L.) breeding lines in Tunisia. *Plant Production Science* 19, 562–573.

Tuinstra, M.R., Soumana, S., Al-Khatib, K., Kapran, I., Toure, A. *et al.* (2009) Efficacy of herbicide seed treatments for controlling *Striga* infestation of sorghum. *Crop Science* 49, 923–929.

Van Ast, A., Bastiaans, L. and Katile, S. (2005) Cultural control measures to diminish sorghum yield loss and parasite success under *Striga hermonthica* infestation. *Crop Protection* 24, 1023–1034.

Van Delft, G.J., Graves, J.D., Fitter, A.H. and Van Ast, A. (2000) Striga seed avoidance by deep planting and no-tillage in sorghum and maize. *International Journal of Pest Management* 46, 251–256.

Van Hezewijk, M.J. and Verkleij, J.A.C. (1996) The effect of nitrogenous compounds on in vitro germination of *Orobanche crenata* Forsk. *Weed Research* 36, 395–404.

Van Mourik, T.A., Bianchi, F.J.J.A., Van Der Werf, W. and Stomph, T.J. (2008) Long-term management of *Striga hermonthica*: strategy evaluation with a spatio-temporal population model. *Weed Research* 48, 329–339.

Vissoh, P.V., Gbèhounou, G., Ahanchede, A., Roling, N.G. and Kuyper, T.W. (2008) Evaluation of integrated crop management strategies employed to cope with *Striga* infestation in permanent land use systems in southern Benin. *International Journal of Pest Management* 54, 197–206.

Vurro, M., Boari, A., Evidente, A., Andolfi, A. and Zermane, N. (2009) Natural metabolites for parasitic weed management. *Pest Management Science* 65, 566–571.

Wang, R. (1986) Current status and perspectives of biological weed control in China. *Chinese Journal of Biological Control* 1, 173–77.

Westwood, J.H. and Foy, C.L. (1999) Influence of nitrogen on germination and early development of broomrape (*Orobanche* spp.). *Weed Science* 47, 2–7.

Yaacoby, T., Goldwasser, Y., Paporish, A. and Ruben, B. (2015) Germination of *Phelipanche aegyptiaca* and *Cuscuta campestris* seeds in composted farm manure. *Crop Protection* 72, 76–82.

Yacoubou, A.M., Zoumarou Wallis, N., Menkir, A., Zinsou, V.A., Onzo, A. *et al.* (2021) Breeding maize (*Zea mays*) for *Striga* resistance: past, current and prospects in sub-Saharan Africa. *Plant Breeding* 140, 195–210.

Yoneyama, K., Xie, X., Kusumoto, D., Sekimoto, H., Sugimoto, Y. *et al.* (2007) Nitrogen deficiency as well as phosphorus deficiency in sorghum promotes the production and exudation of 5-deoxystrigol, the host recognition signal for arbuscular mycorrhizal fungi and root parasites. *Planta* 227, 125–132.

Yoneyama, K., Awad, A.A., Xie, X.N. and Takeuchi, Y. (2010) Strigolactones as germination stimulants for root parasitic plants. *Plant and Cell Physiology* 51, 1095–1103.

Zehhar, N., Labrousse, P., Arnaud, M.C., Boulet, C., Bouya, D. *et al.* (2003) Study of resistance to *Orobanche ramosa* in host (oilseed rape and carrot) and non-host (maize) plants. *European Journal of Plant Pathology* 109, 75–82.

Zwanenburg, B. and Thuring, J.W.J.F. (1997) Synthesis of strigolactones and analogs: a molecular approach to the witchweed problem. *Pure and Applied Chemistry* 69, 651–654.

14 Synthesis and Outlook

Abstract
Forty-nine species of parasitic plants are known to occur in crops in Africa, but they vary widely in terms of their biology, ecology, distribution, host range and estimated impact on African economies. Only seven of these parasitic plant species are widely distributed across the continent. Just four, all of them root parasites, are known to constitute significant weed problems in smallholder cropping systems: *Striga asiatica*, *S. hermonthica*, *S. gesnerioides* and *Rhamphicarpa fistulosa*. However, a range of subregional or more crop-specific parasitic plant species are responsible for current important crop losses or are future threats. Increasing our understanding of the current and projected invasiveness and distribution of these species combined with generating knowledge on their biology, ecology and potentially effective management strategies will prepare African agriculture for future parasitic weed problems.

14.1 Introduction

In this book, we have discussed 49 different parasitic plant species that are known weeds in African agriculture, of which 21 are stem parasites and 28 root parasites. For only a few of these species does a significant body of literature exist with sufficient understanding of their biology, ecology, host interactions, agronomic importance and impact to generate information for their management. The root parasites, in particular *Striga hermonthica*, *S. asiatica*, *S. gesnerioides*, *Alectra vogelii* and *Rhamphicarpa fistulosa*, were subject to the majority of parasitic weed studies with relevance to Africa. The broomrapes have been studied widely as well, but not within the context of African farming systems. The stem parasites with African distributions have received less attention and consequently we know much less about their distribution and impact on African agricultural systems as well as best management strategies for these weeds.

 Parasitic Plants in African Agriculture (L.J. Musselman and J. Rodenburg)

DOI: 10.1079/9781789247657.0014

Table 14.1. Number of countries on the African continent where parasitic weed species have been recorded, depending on extent of collections. From GBIF, 2022.

Parasitic weed species	Number of countries where species recorded					
	East	Central	North	South	West	Total Africa
Stem parasites						
Tapinanthus bangwensis	1	4	–	–	14	19
T. belvisii	–	–	–	–	5	5
Erianthemum dregei	10	2	–	4	1	17
Phragmanthera capitata	2	6	–	1	8	17
P. incana	–	1	–	–	4	5
Viscum cruciatum	–	–	1	–	–	1
V. anceps	1	–	–	1	–	2
V. engleri	2	1	–	–	–	3
V. rotundifolium	2	2	–	5	–	9
Cassytha filiformis	14	6	–	4	14	38
Cuscuta campestris syn. *C. pentagona*[a]	8	2	2	3	3	18
C. epilinum	1	–	–	–	–	1
C. suaveolens	–	–	2	1	–	3
C. hyalina	5	1	2	2	1	11
C. monogyna	–	–	2	–	–	2
C. epithymum	2	–	3	1	–	6
C. pedicellata	–	–	3	–	–	3
C. planiflora	8	4	6	2	1	21
C. australis syn. *C. scandens*	5	2	1	2	6	16
C. chinensis	3	–	1	–	1	5
C. kilimanjari	10	1	1	1	–	13
Number of stem-parasite species per subregion	15	12	11	12	11	21
Root parasites						
Rhamphicarpa fistulosa	12	7	2	4	12	37
R. brevipedicellata	1	–	–	3	–	4

R. capillacea	–	3	–	–	–	3
R. elongata	1	1	1	–	–	3
R. veronicaefolia	2	–	–	–	–	2
Buchnera hispida	12	4	1	3	10	30
Striga asiatica	14	7	2	5	14	42
S. aspera	2	3	1	–	13	19
S. hermonthica	10	6	2	1	13	32
S. gesnerioides	12	5	2	5	11	35
S. forbesii	13	3	1	3	6	26
S. passargei	1	1	1	–	9	12
S. brachycalyx	–	1	–	1	8	10
Alectra vogelii and *A. picta*	8	3	–	4	7	22
A. sessiliflora	14	8	1	4	13	40
A. orobanchoides	9	3	–	5	–	17
Orobanche crenata	–	–	5	–	1	6
O. cernua	–	–	7	–	–	7
O. minor	12	1	5	1	–	19
O. foetida	–	–	3	–	–	3
Phelipanche aegyptiaca	–	–	3	–	–	3
P. ramosa	4	–	5	2	–	11
Thonningia sanguinea	7	7	–	–	13	27
Micrargeria filiformis	9	4	–	–	11	24
Sopubia parviflora	5	4	1	–	10	20
Thesium humile	–	–	5	–	–	5
T. resedoides	1	1	–	3	–	5
Number of root-parasite species per subregion	20	19	18	14	15	27

[a]Recent work has shown these are two distinct species, but they are easily confused and it is likely that some of the reports of *C. campestris* are indeed *C. pentagona*.

14.2 Parasitic Weed Species Distribution

Across the continent, seven of the 49 parasitic weed species are recorded in 30 or more countries (Table 14.1). *Striga asiatica* (42 countries) is the most widely distributed species, followed by *Alectra sessiliflora* (40), *Cassytha filiformis* (38), *Rhamphicarpa fistulosa* (37), *S. gesnerioides* (35), *S. hermonthica* (32) and *Buchnera hispida* (30). Three of these seven species (i.e. *A. sessiliflora, C. filiformis* and *B. hispida*) are not known as serious weed species, showing that a wide distribution does not necessarily render a species a problematic weed. On the other end of the spectrum, 22 of the 49 parasitic weed species are recorded in fewer than ten countries and none of them is known as a problem weed in African agriculture.

The east and centre of the African continent have the most parasitic weed species diversity with respectively 71% and 57% of the stem parasites and 77% and 73% of the root parasites recorded in the database of the Global Biodiversity Information Facility (GBIF, 2022). The highest number of stem-parasitic weed species (12 species) is recorded in South Africa (Table 14.2). Other countries with seven or more stem-parasitic weed species are Tanzania (eight species) and Burundi, Cameroon, Democratic Republic of the Congo (DRC), Kenya and Nigeria (all with seven species). In four countries, including the Central African Republic and South Sudan, stem-parasitic weed species were not recorded, and in 13 counties only one (including the Gambia, Libya, Niger) or two (Burkina Faso, Chad, Comoros, Congo, Eritrea, Guinea-Bissau, Mali, and Tunisia) stem-parasite species were recorded. Countries with the highest number of root-parasitic weed species are Cameroon and Tanzania (15 species); Benin, Burkina Faso, DRC, Ethiopia, Mali, Nigeria and Zambia (14 species); and Angola, Kenya, South Africa and Sudan (13 species) (Table 14.2). For only two countries (the Indian Ocean islands of Mauritius and Seychelles) are there no records of root-parasitic plant species.

Although GBIF database records should provide a good indication of the distribution of species, an analysis per country should be interpreted with caution as the extent of collections included in the database is a function of the intensity of botanical research and this may vary widely across countries. Furthermore, presence of a parasitic plant species in a country does not necessarily mean the species represents a weed problem in that country. First, the GBIF records do not provide information on the abundance of plants of a particular species at a given location. Second, species distribution in a country may not necessarily overlap with crop production areas and if they do, the crop species may not necessarily comprise suitable hosts for the parasitic plant species. Third, the GBIF database includes historic records and, in a few cases, species may have gone extinct since the date of recording. However, records of parasitic plant species in a country provide strong indications that the country harbours suitable environments and there is a high likelihood that the species is present. Presence of a parasitic plant species in each country, even if it is not yet observed in crop production systems, is important information from a weed science and management perspective. Parasitic plant species present in natural vegetation may spread into agricultural areas and become weed problems. The main cause for such spread would be an expansion of arable land.

Table 14.2. Number of stem-parasitic and root-parasitic weed species recorded per country (depending on extent of collections).

Northern Africa	Stem	Root	Eastern Africa	Stem	Root	Central Africa	Stem	Root	Southern Africa	Stem	Root	Western Africa	Stem	Root
Algeria	6	5	Burundi	7	11	Angola	6	13	Botswana	4	9	Benin	6	14
Egypt	3	8	Comoros	2	4	Cameroon	7	15	Eswatini	4	8	Burkina Faso	2	14
Libya	1	3	Djibouti	0	1	Central African Republic	0	9	Lesotho	3	4	Cabo Verde	1	4
Morocco	6	7	Eritrea	2	6	Chad	2	8	Namibia	4	10	Côte d'Ivoire	6	12
Sudan	6	13	Ethiopia	6	14	Congo	2	6	South Africa	12	13	Gambia	1	6
Tunisia	2	5	Kenya	7	13	Democratic Republic of the Congo	7	14				Ghana	6	12
			Madagascar	6	8							Guinea	4	12
			Malawi	6	10	Equatorial Guinea	4	2				Guinea-Bissau	2	9
			Mauritius	3	0	Gabon	3	4				Liberia	3	3
			Mozambique	4	12	Sao Tome and Principe	1	1				Mali	2	14
			Rwanda	4	9							Mauritania	3	3
			Seychelles	0	0							Niger	1	8
			Somalia	4	5							Nigeria	7	14
			South Sudan	0	7							Senegal	4	11
			Tanzania	8	15							Sierra Leone	4	5
			Uganda	6	9							Togo	6	10
			Zambia	4	14									
			Zimbabwe	5	11									

14.3 Impact and Importance of Parasitic Weeds in Africa

A common problem regarding parasitic weeds in African agriculture, is a genuine scarcity of data. For the majority of the 49 species discussed in this book, there is no or very little information on their importance as weeds in African cropping systems (e.g. occurrence, abundance, affected crops, associated yield losses) and there is a scarcity of information on the impact they have on smallholder farmers' livelihoods, African domestic economies and food security. Even for a parasitic weed genus as widely studied as *Striga*, data scarcity exists. Systematic surveys and assessments would be needed, using GIS/RS and modelling tools, to obtain reliable data on the importance of parasitic weeds by crop category. Based on the distribution of parasitic weed species (GBIF, 2022; which depends on the extent of collections of species per country) and their host crop ranges, combined with data on harvested crop area (FAOSTAT, 2022; which depend on reliability of national reports) it is possible to indicate countries and crops that may be under threats of parasitic weed infestation, currently or in the future. But caution should be taken with these indications as they have a low resolution and would need to be verified on the ground.

14.3.1 Stem parasites

14.3.1.1 Mistletoe

Viscum spp., *Tapinanthus* spp., *Phragmanthera* spp. and *Erianthemum dregei* can parasitize tree crops and have a broad host range. They could therefore be encountered as weeds in fruit trees, nut trees and plantation cash crops (cacao, coffee, rubber, tea). Important tree-crop production countries where mistletoes are observed are Tanzania, Uganda, Kenya, Ethiopia, DRC, Mozambique and South Africa (*Erianthemum dregei*); Tanzania, Malawi, Morocco, Mozambique and South Africa (*Viscum* spp.); Guinea-Bissau, Burkina Faso (*Tapinanthus* spp.); Côte d'Ivoire, Nigeria, Ghana, Benin, Cameroon, Guinea, Sierra Leone, Liberia, Togo (*Tapinanthus* spp. and *Phragmanthera* spp.); and DRC (*Phragmanthera* spp.). Among the listed countries are some world-leading cacao, coffee and rubber producers, and mistletoes may therefore particularly affect the economies of these countries.

14.3.1.2 Love vine

Cassytha filiformis is one of the most widespread parasitic weed species in tree crops. As well as being a risk in the countries listed for mistletoe above, love vine may threaten economically important tree production systems in Madagascar (including coffee, clove and cinnamon, cashew and cacao), as well as in Mali (e.g. karité/shea nut, cashew, mango) and Sudan (e.g. citrus and mango).

14.3.1.3 Dodder

Cuscuta spp. have very broad host ranges (including tree crops, pulses, root and tuber crops, forage crops, fibre crops and vegetables) and distributions across the continent. This complicates prioritization based on species distribution and crop-area data. Based on published data, tea, coffee and mango production systems in Kenya are under threat (Masanga *et al.*, 2021). All other tree-crop-producing countries where dodder is found would need to be prepared for increasing future infestations too (Table 14.3). Based on data on area under fruit and vegetable production and the percentage production area of total land area of a country (Table 14.4), Nigeria, Rwanda and Burundi would need to be vigilant.

14.3.2 Root parasites

14.3.2.1 Rice vampire weed

Rhamphicarpa fistulosa is mainly observed in rice but can also infest sorghum and maize (Ouédraogo *et al.*, 1999; Rodenburg *et al.*, 2015). Although the species has broad ecological amplitude, the ideal habitats for the parasite are the wet hydromorphic zones of rainfed lowlands (often referred to as 'inland valleys') that are only temporarily flooded during peaks in the rainy season (Rodenburg *et al.*, 2015). In areas where floods are increasing due to climate change, this parasitic weed may become more prevalent (Kabiri *et al.*, 2015). Countries under serious current and future threat because of the overlap between the species distribution, a high density of rice-producing inland valleys and our field observations are Burkina Faso, Cameroon, Côte d'Ivoire, Guinea, Mali, Nigeria and Sierra Leone in Western Africa, and Madagascar, Tanzania and Uganda in Eastern Africa (Rodenburg *et al.*, 2015).

14.3.2.2 Cereal witchweed

Striga asiatica, *S. aspera* and *S. hermonthica* are among the most widely distributed and most serious parasitic weeds in Africa. An important reason this is a continent-wide concern is that these parasites attack the main cereal food crops. An estimated 35 million ha under maize, sorghum, millet and rice production are infested by witchweed (see Chapter 7, this volume) and this directly affects food security in the region. The most-impacted countries based on their share of land under cereal production (Table 14.4), the distribution of these species and our own observations are Burkina Faso, Ethiopia, Guinea, Kenya, Malawi, Mali, Niger, Nigeria, Senegal, Sudan, Tanzania and Uganda as well as the smaller countries such as the Gambia, Benin, Burundi and Togo.

14.3.2.3 Alectra and cowpea witchweed

Alectra vogelii and *Striga gesnerioides* affect pulses, primarily cowpea. Based on the distribution of these species, the literature, field observations and figures regarding the importance of pulses in terms of area under production (Table 14.4), these parasites are a particular concern to countries such as Burkina Faso, Ethiopia, Kenya, Malawi, Niger (only *S. gesnerioides*), Nigeria,

Table 14.3. Tree-crop production area (expressed in 1000 ha) and share (% production area of total land area) per tree-crop category[a] for the 34 most important tree-crop producing countries in Africa (based on total area under production and share of total land area; sorted on importance). Bold figures indicate the most important countries of Africa for a given crop category. Data are for 2020, obtained from FAOSTAT (2022).

Country	Fruits	%	Nuts	%	Oils	%	Plantation	%	SS&A	%
Côte d'Ivoire	**186**	**0.6**	**2086**	**6.5**	**383**	**1.2**	**6110**	**18.9**	**85**	**0.3**
Tunisia	**196**	**1.2**	**199**	**1.2**	**3643**	**22.3**	0	0.0	0	0.0
Nigeria	**973**	**1.1**	**175**	0.2	**4079**	**4.4**	**1630**	**1.8**	**256**	**0.3**
Ghana	35	0.1	**243**	**1.0**	**390**	**1.6**	**1509**	**6.3**	**86**	**0.4**
Morocco	**369**	**0.8**	**216**	0.5	**1069**	**2.4**	0	0.0	0	0.0
Burundi	0	0.0	**278**	**10.0**	9	0.3	29	**1.1**	0	0.0
Tanzania	97	0.1	**1423**	**1.5**	5	0.0	**232**	0.2	**8**	0.0
Benin	10	0.1	**546**	**4.8**	43	**0.4**	0	0.0	**2**	**0.0**
Cameroon	19	0.0	3	0.0	**168**	0.4	**869**	**1.8**	**108**	**0.2**
Guinea-Bissau	3	0.1	**303**	**8.4**	10	0.3	0	0.0	0	0.0
Uganda	0	0.0	0	0.0	0	0.0	**637**	**2.6**	0	0.0
Guinea	**135**	**0.6**	26	0.1	**317**	**1.3**	129	0.5	0	0.0
Sao Tome and Principe	0	0.0	0	0.2	2	**1.8**	31	**32.0**	0	**0.2**
Ethiopia	63	0.1	58	0.1	0	0.0	**866**	0.8	0	0.0
Comoros	0	0.0	28	**15.0**	0	0.0	1	0.6	**8**	**4.4**
Sierra Leone	25	**0.4**	5	0.1	37	**0.5**	134	**1.9**	**28**	**0.4**
Kenya	**123**	**0.2**	112	0.2	0	0.0	**389**	0.7	**2**	0.0
Liberia	2	0.0	5	0.0	18	0.2	**177**	**1.6**	0	0.0
Algeria	**394**	0.2	35	0.0	**439**	0.2	0	0.0	0	0.0
Egypt	**446**	**0.4**	6	0.0	101	0.1	0	0.0	0	0.0
Togo	3	0.1	9	0.2	68	**1.2**	49	0.9	0	0.0
Burkina Faso	2	0.0	152	**0.6**	100	0.4	0	0.0	0	0.0
Madagascar	107	0.2	64	0.1	2	0.0	91	0.2	**76**	**0.1**
Malawi	119	**1.0**	9	0.1	0	0.0	23	0.2	0	0.0
Democratic Republic of the Congo	39	0.0	0	0.0	**329**	0.1	**163**	0.1	0	0.0
Mozambique	12	0.0	**241**	0.3	0	0.0	46	0.1	0	0.0
Libya	61	0.0	60	0.0	**239**	0.1	0	0.0	0	0.0

Rwanda	4	0.2	0	0.0	0	0.0	38	**1.5**	0	0.0
Mali	120	0.1	42	0.0	67	0.1	0	0.0	0	0.0
Equatorial Guinea	0	0.0	4	0.2	4	0.1	21	0.8	0	0.0
South Africa	**168**	0.1	17	0.0	0	0.0	1	0.0	0	0.0
Senegal	26	0.1	25	0.1	11	0.1	0	0.0	0	0.0
Sudan	**144**	0.1	0	0.0	0	0.0	0	0.0	0	0.0
Angola	39	0.0	3	0.0	23	0.0	52	0.0	0	0.0

[a]Tree-crop category: **fruits** (fruit trees, including apple, apricot, avocado, cashew, cherry, date, guava, fig, kiwi, lemon, mango and mangosteen, orange and clementine, pomelo and grapefruit, peach and nectarine, pear, plum and sloe, quince, tangerine and mandarin); **nuts** (nut trees, including almond, cashew nut, coconut, hazelnut, pistachio, tung nut, walnut); **oils** (oil trees, including karité/shea nut, oil palm, olive); **plantation** (plantation cash crop trees, including cacao, coffee, rubber, tea); and **SS&A** (stimulants, spices and aromatics trees, including cinnamon, clove, kola nut, maté, nutmeg).

Table 14.4. Field-crop production area (expressed in 1000 ha) and share (% production area of total land area) per field-crop category[a] for the 34 most important field-crop producing countries in Africa (based on total area under production and share of total land area; sorted on importance). Bold figures indicate the most important countries of Africa for a given crop category. Data are for 2020, obtained from FAOSTAT (2022).

Country	Cereals	%	S. cane	%	Pulses	%	RT&P	%	F&V	%	SS&A	%	Oilseed	%	Fibre	%
Nigeria	**20,394**	**22**	**86**	0.1	**5,489**	**5.9**	**17,174**	**18.6**	**4,375**	**4.7**	**236**	**0.3**	**5,595**	**6.1**	**375**	**0.4**
Niger	**10,494**	8	10	0.0	**5,893**	**4.7**	37	0.0	273	0.2	24	0.0	**1,081**	0.9	10	0.0
Tanzania	**6,894**	7	**52**	0.1	**1,387**	1.5	**2,076**	2.2	**788**	0.8	**131**	0.1	**3,066**	**3.2**	**547**	**0.6**
Burkina Faso	**4,377**	**16**	5	0.0	**1,489**	**5.5**	23	0.1	46	0.2	6	0.0	**930**	**3.4**	**647**	**2.4**
Sudan	**8,571**	5	**69**	0.0	**1,015**	0.5	144	0.1	**412**	0.2	14	0.0	**8,987**	**4.8**	**202**	0.1
Uganda	1,532	6	**81**	**0.3**	577	2.4	**3,451**	**14.3**	247	1.0	28	0.1	**875**	**3.6**	94	**0.4**
Ethiopia	**10,572**	10	30	0.0	**1,679**	1.5	245	0.2	**415**	0.4	**592**	**0.5**	783	0.7	**399**	**0.4**
Rwanda	516	**20**	8	**0.3**	715	**27.1**	618	**23.5**	290	**11.0**	6	**0.2**	36	1.4	0	0.0
Malawi	2,010	**17**	28	**0.2**	**964**	**8.1**	**656**	**5.5**	220	**1.9**	**99**	**0.8**	431	**3.6**	81	**0.7**
Ghana	2,374	10	6	0.0	657	2.8	**2,100**	**8.8**	102	0.4	40	0.2	330	1.4	15	0.1
Burundi	355	**13**	3	0.1	756	**27.2**	475	**17.1**	241	**8.6**	2	0.1	15	0.5	2	0.1
Guinea	**3,645**	**15**	6	0.0	62	0.3	491	2.0	147	0.6	2	0.0	**986**	**4.0**	46	0.2
Benin	1,538	**14**	1	0.0	388	**3.4**	586	5.2	132	**1.2**	**53**	**0.5**	186	**1.7**	**620**	**5.5**
Togo	1,158	**20**	0	0.0	444	**7.8**	406	**7.1**	36	0.6	13	**0.2**	68	1.2	158	**2.8**
Côte d'Ivoire	1,347	4	26	0.1	68	0.2	**2,974**	**9.2**	182	0.6	46	0.1	187	0.6	**440**	**1.4**
Senegal	1,999	10	12	0.1	291	1.5	105	0.5	187	0.9	1	0.0	**1,280**	**6.5**	18	0.1
Cameroon	2,247	5	**137**	**0.3**	601	1.3	**1,058**	2.2	**874**	**1.8**	52	0.1	625	1.3	**250**	**0.5**
DRC	**4,140**	2	47	0.0	844	0.4	**6,443**	2.7	**347**	0.1	18	0.0	611	0.3	80	0.0
Morocco	**4,447**	10	12	0.0	415	0.9	116	0.3	195	0.4	42	0.1	36	0.1	5	0.0
Mozambique	2,899	4	47	0.1	**1,607**	2.0	**903**	1.1	213	0.3	**133**	0.2	**958**	1.2	144	0.2
Kenya	2,703	5	**90**	0.2	**1,743**	3.0	298	0.5	**302**	0.5	20	0.0	159	0.3	39	0.1
Mali	**6,140**	5	5	0.0	488	0.4	66	0.1	237	0.2	20	0.0	552	0.4	**167**	0.1
Egypt	3,607	4	**136**	0.1	135	0.1	460	0.5	**731**	0.7	**124**	0.1	105	0.1	75	0.1
Angola	2,678	2	20	0.0	**955**	0.8	**1,169**	0.9	**417**	0.3	3	0.0	403	0.3	5	0.0
South Africa	3,409	3	**283**	**0.2**	782	0.6	99	0.1	247	0.2	16	0.0	627	0.5	49	0.0
Chad	3,324	3	5	0.0	277	0.2	126	0.1	29	0.0	0	0.0	**1,185**	0.9	**220**	0.2
Sierra Leone	717	10	1	0.0	124	1.7	150	2.1	68	0.9	**78**	**1.1**	49	0.7	0	0.0

Madagascar	1,805	3	**95**	**0.2**	112	0.2	544	0.9	193	0.3	**87**	0.1	85	0.1	37	0.1
Tunisia	1,179	7	0	0.0	113	0.7	28	0.2	145	0.9	48	**0.3**	13	0.1	3	0.0
Gambia	233	**22**	0	0.0	12	1.1	3	0.3	4	0.3	0	0.0	118	**11.0**	2	0.1
Zimbabwe	1,488	4	47	0.1	145	0.4	68	0.2	56	0.1	**108**	**0.3**	200	0.5	120	0.3
Zambia	1,486	2	47	0.1	277	0.4	194	0.3	82	0.1	16	0.0	287	0.4	68	0.1
Algeria	2,890	1	0	0.0	189	0.1	149	0.1	**339**	0.1	43	0.0	16	0.0	0	0.0
South Sudan	756	1	0	0.0	37	0.1	103	0.2	142	0.2	4	0.0	777	1.3	0	0.0

[a]Field-crop categories: **cereals** (cereal crops, including barley, fonio, maize, millet, oat, rice, rye, sorghum, triticale, wheat, teff); **s. cane** (sugarcane); **pulses** (pulses or leguminous crops, including bambara bean, bean, broad bean and horse bean, cowpea, chickpea, lentils, lupin, peas, soybean, string bean, vetch); **RT&P** (root and tuber crops and plantains, including cassava, enset, plantain, potato, sweet potato, sugar beet, taro, yam); **F&V** (fruits and vegetables, including artichoke, asparagus, banana, berries, cabbage, carrot and turnip, cauliflower and broccoli, cucumber and gherkin, aubergine, leek and other alliaceous vegetables, lettuce and chicory, melon, okra, onion and shallot, papaya, pineapple, pumpkin (squash and gourd), spinach, strawberry, tomato, watermelon); **SS&A** (stimulants, spices and aromatics, including anise, chilli and pepper, chicory root, coriander, cumin and fennel, garlic, ginger, hops, tobacco, peppermint, vanilla); **oilseed** (oilseed crops, including castor oil, groundnut, linseed, melonseed, mustard, rape or colza, safflower, sesame, sunflower); and **fibre** (fibre crops, including flax, jute, kenaf, cotton, sisal).
DRC = Democratic Republic of the Congo.

Benin and Burundi. Important groundnut-growing countries (in the category 'oilseed crops' in Table 14.4) not listed above such as Chad (only *S. gesnerioides*), Mali and Senegal are also under threat.

14.3.2.4 Broomrape

Orobanche minor, O. crenata, O. cernua, O. foetida, Phelipanche aegyptiaca and *P. ramosa* are mainly restricted in their distributions to Northern Africa, perhaps with the exception of *O. minor* and *P. ramosa*, which are also widely distributed in Eastern and Southern Africa (Table 14.1). Based on species distributions and importance of crops, the production of potatoes and vegetables such as tomato and eggplants are threatened by *P. ramosa* in Tanzania and Kenya and northern countries such as Sudan, Ethiopia, Egypt and Algeria. *Orobanche minor* is also distributed in these countries and may also cause reductions in vegetable production in countries such as Malawi, Rwanda, Uganda and Tunisia. *Orobanche crenata* and *O. foetida* may cause production losses in pulses in Morocco and Tunisia.

14.3.2.5 Thonningia

Thonningia sanguinea is widely distributed, in a belt from Western Africa to Madagascar, and is mainly a problem in plantation tree crops such as rubber, coffee and cacao. Due to the species distribution and the economic importance of plantation tree crops, *T. sanguinea* is of particular concern for countries such as Sierra Leone, Liberia, Côte d'Ivoire, Ghana, Nigeria, Cameroon and Uganda.

14.4 Future Parasitic Weed Problems

14.4.1 Major parasitic weeds becoming more widespread or impactful

Distributions and therefore problems of established parasitic weeds in Africa have been projected to increase by several studies. *Striga* spp. and *Rhamphicarpa fistulosa*, for instance, are both predicted to increase, based on evidence of recent trends. The area of sorghum, millet, maize and rice production systems infested by *Striga hermonthica* have been observed to increase by 1.2% in Ghana (Aflakpui *et al.*, 2008), 1.6% in Nigeria (Lagoke *et al.*, 1991, compared with Dugje *et al.*, 2006) and 2.2% in Côte d'Ivoire (Kouakou *et al.*, 2015). The proportion of inland valleys (rainfed lowland areas) infested by *Rhamphicarpa fistulosa* has increased from 33% to 55% within a decade, an annual increase of 2.2% (Gbèhounou and Assigbé, 2003, compared with Rodenburg *et al.*, 2011).

Parasitic weed species may spread to areas that would inherently have suitable environmental conditions, or to areas where environmental conditions change in their favour, for instance due to climate change. Hättenschwiler and Zumbrunn (2006) have shown that hemiparasitic plant abundance may increase with incremental atmospheric CO_2 concentrations, although this was demonstrated in a European mountain range at height comprising very specific growing conditions. Grenz and Sauerborn (2007) concluded that climates in Western Africa would favour *Orobanche crenata*. Also, Mohamed *et al.* (2006) predicted based on ecological niche modelling that species of the Orobanchaceae are likely to be favoured by climate changes. Using similar methods, Kimathi *et al.* (2022) recently predicted an increase of 19%–53% of the suitable area for *Striga hermonthica* in western Kenya between 2021 and 2050, due to climate change (mainly changes in rainfall and temperature). Based on such types of analysis, Masanga *et al.* (2021) assessed large areas in Eastern Africa as highly suitable for *Cuscuta* spp. Given that these species have a wide host range (including important export crops such as coffee and tea), there is a high potential for them to spread further than their current distribution.

An increase in temperatures may also mean that crops grown at higher altitudes or greater latitudes may become more prone to new infestations if parasitic weed species spread into their growing environments. This may for instance lead to *Striga* spp. infestations in wheat, barley or teff production areas.

An actual increase in infested areas would still require introductions of the seed from other areas, for instance through soil erosion, water or wind distribution (some of which may be aggravated by climate changes), or introductions by humans, through seeds or other means. The latter seems to be the case with *Cuscuta reflexa* based on observations of recent invasions by this dodder species in two areas in Kenya, reported by Masanga *et al.* (2021). Hence, following preventive, sanitary measures and monitoring of new infestations will become more important than ever.

New parasitic weeds may also emerge when natural vegetation is transformed into agricultural land. When the crops grown on newly cleared land comprise suitable hosts of parasitic plant species that occurred in the natural vegetation, these parasitic plants may become very successful and turn into a weed due to the sudden increase in density of suitable host plants. This is one of the scenarios behind the emergence of *Rhamphicarpa fistulosa* in rainfed lowland rice witnessed in the past decades.

Increased drought occurrences because of climate change may have a more pronounced negative effect on crops that are already infected by parasitic weeds. Future environmental changes may however not just mean bad news when it comes to parasitic weeds. Increasing atmospheric CO_2 levels may benefit the tolerance of parasitic-weed-infected host plants with C_3 photosynthetic pathways such as rice (Watling and Press, 2000). Host plants may reduce their transpiration rates and may therefore become less drought sensitive when they are parasitized by parasitic weeds with intrinsic high transpiration rates such as *Striga* spp.

14.4.2 Minor parasitic weeds becoming major ones

In this section we will consider root parasites that affect annual field crops, as such parasitic weeds are no doubt the most important in Africa and field-crop production systems are likely to expand in area and intensify in crop frequency in the future in response to increasing demands for food. With intensification, weed problems may become more urgent.

Of the roughly 150 species of root parasites in Africa, all but two of which are herbaceous, which can become crop parasites? This is perhaps one of the most intriguing overarching parasitic weed science questions deserving attention. It is unclear why among comparable facultative parasitic plants one species, such as *Rhamphicarpa fistulosa*, is an agricultural threat, whereas other species from the same genus (i.e. *R. brevipedicellata, R. capillacea, R. elongata, R. veronicaefolia*) remain isolated in distribution and are not reported as weeds. Another example is *Buchnera hispida*, which despite its wide distribution and a similar host range remains an insignificant problem. Similar questions can be raised with respect to *Striga* spp. In southern Tanzania (Kyela District) we observed two species of *Striga* (*S. asiatica* and *S. forbesii*) seemingly sharing the same environments. However, whereas *S. asiatica* was rampant in the upland rice crops in that area, *S. forbesii* was only sporadically observed in the field margins. Why are *S. hermonthica*, and to a lesser extent *S. aspera*, important parasitic weeds in northern Côte d'Ivoire (Ferkessedougou, Korhogo, Boundiali) in rice, sorghum, maize and millet (only *S. hermonthica*) whereas the local ecotype of *S. asiatica*, as well as the close relative *S. brachycalyx*, both present in the same environments as *S. hermonthica*, are not doing much harm (as shown by Johnson *et al.*, 1997)? Does environmental plasticity play a role, or rather host specificity and/or differences in virulence between ecotypes? There are certainly indications of that. Is it associated with prolificity, with seed dormancy and germination biology, or does it have to do with species-specific responses to agricultural inputs? Further research is needed.

As the above examples show, many of the indigenous parasites of Africa pose little or no threat to agriculture. On the other hand, agriculturalists should be aware of the potential of native parasites becoming a problem. There are several records of seemingly non-pathogenic root parasites becoming serious problems. The best African example is *Rhamphicarpa fistulosa*, which was previously known as a weed of minor but potential importance (Raynal-Roques, 1994) but became an important weed of lowland rice in the 25 years after that prediction (Chapter 5, this volume; Rodenburg *et al.*, 2015, 2016). Related to the above question is, therefore, how can we predict which of the native African parasites as yet unknown as important weeds might become agricultural problems in the future?

Raynal-Roques (1994) suggested some criteria for identifying potential parasitic weeds: (i) being related to current parasitic weed species; (ii) having a broad host range; (iii) being adapted to different soil moisture regimes; and (iv) known to attack food crops or relatives of food crops. This important question could be addressed by knowing which parasites have been documented parasitizing crop plants. This would ideally involve documenting host preferences of these parasites. Knowing host ranges, trends in crop production area and studies on effects of environmental drivers could be used as predictors. One of the reasons parasitic weeds (e.g. *R. fistulosa* but also *S. asiatica* and *S. hermonthica*) have been important in rice systems in the last three decades, for instance, is the increasing area under rainfed production in sub-Saharan Africa, which in turn was partly driven by the generation and release of the popular New Rice for Africa (NERICA) varieties with broad adaptation to rainfed rice-growing environments (Rodenburg *et al.*, 2016).

Part of the reason certain native parasites are not yet known as problematic weeds may be that they are overlooked. Again, our experience with *Rhamphicarpa fistulosa* supports this scenario. It was not well known by weed scientists and extension services and therefore not noticed (see Chapter 5, this volume; Rodenburg *et al.*, 2015). Once identified in one place, it was soon found in many more places. Two native parasites with the potential to be next discovered as weeds are *Micrargeria filiformis* and *Sopubia parviflora*. These species are relatively widespread in sub-Saharan Africa, in particular *M. filiformis*, and they are found in similar environments as *R. fistulosa*, for example wetlands and swamps. Their habit is unremarkable with narrow leaves and small flowers and they are thus easily overlooked by farmers, extension agents and researchers. Agricultural workers should be aware of these native parasites favouring subhumid to humid savannahs and therefore being potential problems in rainfed lowland rice cultivation. They are both annuals and produce prodigious amounts of wind-dispersed seeds.

The most likely native parasites that could invade cultivated fields have host ranges restricted to herbaceous hosts and exhibit weedy tendencies (i.e. they are agrestal). As we report, *Orobanche foetida*, native to the Maghreb, is now a serious problem on broadbeans in Tunisia. That genus has few native species in tropical regions. On the other hand, hemiparasitic Orobanchaceae including species in the genera *Bartsia, Cycnium, Harveya* and *Melasma* deserve attention. Species of the genus *Thesium* (Thesiaceae or Santalaceae), more specifically *T. humile* (in Northern Africa) and *T. resedoides* (in Southern Africa), also need to be monitored closely.

More attention to these types of weeds in training curricula for agricultural extension services has been proposed to generate more awareness and preparation for dealing with emerging parasitic weed problems in African farming systems (Schut *et al.*, 2015).

14.5 Concluding Remarks

Parasitic plants in Africa comprise a hugely diverse group of angiosperms in terms of biology, ecology and habit. They infect a wide diversity of hosts and there is virtually no crop species not threatened by any parasitic weed. The effective management of these parasitic weeds requires a good understanding of their distribution, diaspore, host interactions, biology and ecology. Obviously, this requires resources, but we believe this information will form the precondition for the design of effective management strategies, preferably based on integrated measures.

To address the heterogeneity of the parasite and the affected African farming systems effectively and durably, research on parasitic weed management needs to generate a 'basket of options'. This will enable the farmer to apply a locally adapted integrated parasitic weed management strategy. That 'basket of options' has been partially filled in recent years but is by no means complete. The parasitic weed problem, like many other agricultural pest problems, continues to evolve under changing climates and farm management. There is therefore a need for continuous innovation in parasitic weed management.

Apart from investments in research, awareness needs to be raised and communication improved between stakeholders at different levels, such as farmers, extension services, crop protection services and an interdisciplinary group of researchers. This would aid in timely identification of problem species and regions, preventing further spread, as well as in the design and dissemination of good management strategies. All this requires good organization and above all good education. Parasitic plant botany and weed ecology and management would need to be integral parts of curricula at agricultural colleges and universities. This would equip agricultural professionals of the future and enable them to recognize emerging species early and start containment and management programmes. This is becoming even more important with changes in climate, making crop production an even riskier and more complicated endeavour and parasitic weeds an even more challenging constraint.

References

Aflakpui, G.K.S., Bolfrey-Arku, G.E.K., Anchirinah, V.M., Manu-Aduening, J.A. and Adu-Tutu, K.O. (2008) Incidence and severity of *Striga* spp. in the coastal savanna zone of Ghana. Results and implications of a formal survey. *Outlook on Agriculture* 37, 219–224.

Dugje, I.Y., Kamara, A.Y. and Omoigui, L.O. (2006) Infestation of crop fields by *Striga* species in the savanna zones of northeast Nigeria. *Agriculture, Ecosystems & Environment* 116, 251–254.

FAOSTAT (2022) Food and Agriculture Organization of the United Nations. Available at: www.fao.org/faostat/en (accessed 21 October 2022).

Gbèhounou, G. and Assigbé, P. (2003) *Rhamphicarpa fistulosa* (Hochst.) Benth. (Scrophulariaceae): new pest on lowland rice in Benin. Results of a survey and immediate control possibilities. *Annales de Sciences Agronomique Bénin* 4, 89–103.

GBIF (2022) Global Biodiversity Information Facility. Available at: www.gbif.org (accessed 18 May 2023).

Grenz, J.H. and Sauerborn, J. (2007) Mechanisms limiting the geographic range of the parasitic weed *Orobanche crenata*. *Agriculture, Ecosystems & Environment* 122, 275–281.

Hättenschwiler, S. and Zumbrunn, T. (2006) Hemiparasite abundance in an alpine treeline ecotone increases in response to atmospheric CO_2 enrichment. *Oecologica* 147, 47–52.

Johnson, D.E., Riches, C.R., Diallo, R. and Jones, M.J. (1997) *Striga* on rice in West Africa; crop host range and the potential of host resistance. *Crop Protection* 16, 153–157.

Kabiri, S., Rodenburg, J., Kayeke, J., Van Ast, A., Makokha, D.W. *et al.* (2015) Can the parasitic weeds *Striga asiatica* and *Rhamphicarpa fistulosa* co-occur in rain-fed rice? *Weed Research* 55, 145–154.

Kimathi, E., Abdel-Rahman, E.M., Lukhoba, C., Ndambi, A., Mudereri, B.T. *et al.* (2022) Ecological determinants and risk areas of *Striga hermonthica* infestation in western Kenya under changing climate. *Weed Research* 63, 45–56.

Kouakou, C.K., Akanvou, L., Bi, I.A.Z., Akanvou, R. and N'da, H.A. (2015) *Striga* species distribution and infestation in cereal food crops of northern Côte d'Ivoire. *Cahiers Agricultures* 24, 37–46.

Lagoke, S.T.O., Parkinson, V. and Agunbiade, R.M. (1991) Parasitic weeds and control methods in Africa. In: Kim, S.K. (ed.) *Combating Striga in Africa. Proceedings of the International Workshop Held in Ibadan, Nigeria, 22–24 August 1988*. International Institute of Tropical Agriculture, Ibadan, Nigeria, pp. 3–14.

Masanga, J., Mwangi, B.N., Kibet, W., Sagero, P., Wamalwa, M. *et al.* (2021) Physiological and ecological warnings that dodders pose an exigent threat to farmlands in Eastern Africa. *Plant Physiology* 185, 1457–1467.

Mohamed, K.I., Papes, M., Williams, R., Benz, B.W. and Peterson, A.T. (2006) Global invasive potential of 10 parasitic witchweeds and related Orobanchaceae. *Ambio* 35, 281–288.

Ouédraogo, O., Neumann, U., Raynal Roques, A., Sallé, G., Tuquet, C. *et al.* (1999) New insights concerning the ecology and the biology of *Rhamphicarpa fistulosa* (Scrophulariaceae). *Weed Research* 39, 159–169.

Raynal-Roques, A. (1994) Major, minor and potential parasitic weeds in semi-arid tropical Africa: the example of Scrophulariaceae. In: Pieterse, A.H., Verkleij, J.A.C. and Ter Borg, S.J. (eds) *Proceedings of the Third International Workshop on Orobanche and Related Striga Research.* Royal Tropical Institute, Amsterdam, pp. 400–405.

Rodenburg, J., Zossou-Kouderin, N., Gbèhounou, G., Ahanchede, A., Touré, A. *et al.* (2011) *Rhamphicarpa fistulosa*, a parasitic weed threatening rain-fed lowland rice production in sub-Saharan Africa – a case study from Benin. *Crop Protection* 30, 1306–1314.

Rodenburg, J., Morawetz, J.J. and Bastiaans, L. (2015) *Rhamphicarpa fistulosa*, a widespread facultative hemi-parasitic weed, threatening rice production in Africa. *Weed Research* 55, 118–131.

Rodenburg, J., Demont, M., Zwart, S.J., and Bastiaans, L. (2016) Parasitic weed incidence and related economic losses in rice in Africa. *Agriculture, Ecosystems & Environment* 235, 306–317.

Schut, M., Rodenburg, J., Klerkx, L., Hinnou, L.C., Kayeke, J. *et al.* (2015) Participatory appraisal of institutional and political constraints and opportunities for innovation to address parasitic weeds in rice. *Crop Protection* 74, 158–170.

Watling, J.R. and Press, M.C. (2000) Infection with the parasitic angiosperm *Striga hermonthica* influences the response of the C_3 cereal *Oryza sativa* to elevated CO_2. *Global Change Biology* 6, 919–930.

Appendix List of Scientific Names of Crops

arugula	*Eruca sativa*
asparagus	*Asparagus officinalis*
aubergine (eggplant)	*Solanum melongena*
avocado	*Persea americana*
basil	*Ocimum basilicum*
bean, lablab	*Dolichos lablab*
cabbage	*Brassica oleracea*
cacao	*Theobroma cacao*
carrot	*Daucus carota*
cashew	*Anacardium occidentale*
cassava	*Manihot esculenta*
cauliflower	*Brassica oleracea* var. *botrytis*
chickpea	*Cicer arietinum*
clover (berseem)	*Trifolium* spp.
coconut	*Cocos nucifera*
coffee	*Coffea arabica, C. robusta*
cola	*Cola nitida*
cowpea	*Vigna unguiculata*
cumin	*Cumin cyminum*
faba bean	*Vicia fava*
flax	*Linum usitatissimum*
fonio	*Digitaria exilis*
groundnut	*Arachis hypogaea*
guava	*Pisidium guajava*
hemp	*Cannabis sativa*
leek	*Allium porrum*
lemon	*Citrus limon*
lentil	*Lens culinaris*

DOI: 10.1079/9781789247657.appx

linseed	*Linum usitatissimum*
lucerne (alfalfa)	*Medicago sativa*
maize	*Zea mays*
mallow, jute (tossa jute)	*Corchorus olitorius*
mango	*Mangifera indica*
marula	*Sclerocarya birrea*
millet, pearl	*Pennisetum glaucum*
mulberry	*Morus alba*
olive	*Olea europaea*
onion	*Allium cepa*
orange, sweet	*Citrus × sinensis*
palm, oil	*Elaeis guineensis*
peach	*Prunus persica*
pecan	*Carya illinoensis*
pigeon pea	*Cajanus cajan*
potato	*Solanum tuberosum*
rice	*Oryza sativa, O. glaberrima*
rubber	*Helvea brasiliensis*
sesame	*Sesamum indicum*
shea nut (karité)	*Vitellaria paradoxa*
sorghum	*Sorghum bicolor, S. vulgare*
soybean	*Glycine max*
sugarcane	*Saccharum officinarum*
sunflower	*Helianthus annuus*
tea	*Camellia sinensis*
teak	*Tectona grandis*
teff	*Eragrostis teff*
tobacco	*Nicotiana tabacum*
tomato	*Solanum lycopersicum*
vetch	*Vicia sativa*

General index

Note: page numbers in italic denote tables and figures.

Index of Crops

Note: page numbers in italic denote tables and figures.